00-CTS-13-ET

USE OF ENHANCED BIOLOGICAL PHOSPHORUS REMOVAL FOR TREATING NUTRIENT-DEFICIENT WASTEWATER

David I. Jenkins, Ph.D.
Willie F. Harper, Jr., Ph.D., P.E.
University of California, Berkeley

2003

Water Environment Research Foundation

The Water Environment Research Foundation, a not-for-profit organization, funds and manages water quality research for its subscribers through a diverse public-private partnership between municipal utilities, corporations, academia, industry, and the federal government. WERF subscribers include municipal and regional water and wastewater utilities, industrial corporations, environmental engineering firms, and others that share a commitment to cost-effective water quality solutions. WERF is dedicated to advancing science and technology addressing water quality issues as they impact water resources, the atmosphere, the lands, and quality of life.

For more information, contact:
Water Environment Research Foundation
635 Slaters Lane, Suite 300
Alexandria, VA 22314-1177
Tel: (703) 684-2470
Fax: (703) 299-0742
www.werf.org
werf@werf.org

This report was co-published by the following organization. For non-subscriber sales information, contact:

IWA Publishing
Alliance House, 12 Caxton Street
London SW1H 0QS, United Kingdom
Tel: +44 (0) 20 7654 5500
Fax: +44 (0) 20 7654 5555
www.iwapublishing.com
publications@iwap.co.uk

Library of Congress Catalog Card Number: 2003109315
Printed in the United States of America

IWAP ISBN: 1-84339-685-8

This report was prepared by the organization(s) named below as an account of work sponsored by the Water Environment Research Foundation (WERF). Neither WERF, members of WERF, the organization(s) named below, nor any person acting on their behalf: (a) makes any warranty, express or implied, with respect to the use of any information, apparatus, method, or process disclosed in this report or that such use may not infringe on privately owned rights; or (b) assumes any liabilities with respect to the use of, or for damages resulting from the use of, any information, apparatus, method, or process disclosed in this report.

University of California, Berkeley

This document was reviewed by a panel of independent experts selected by WERF. Mention of trade names or commercial products does not constitute WERF endorsement or recommendations for use. Similarly, omission of products or trade names indicates nothing concerning WERF's positions regarding product effectiveness or applicability.

ACKNOWLEDGMENTS

The authors thank Ilana Aldor, Trina McMahon, How Ng, Ruth Richardson, Genene Salmon, Andrew Schuler, Natasha Sokolovskaya, Eleanor Wozei, and Susan Yilmaz for assistance in carrying out this research.

Report Preparation

Principal Investigator:
David I. Jenkins, Ph.D.
University of California, Berkeley

Project Team
Willie F. Harper, Jr., Ph.D., P.E.
University of California, Berkeley
(now currently affiliated with Auburn University)

Project Subcommittee

Glen Daigger, Ph.D, P.E., DEE
CH2M HILL, Inc.

H. David Stensel, Ph.D, P.E.
University of Washington

Water Environment Research Foundation Staff

Director of Research: Daniel M. Woltering, Ph.D.
Director of Technical Services: Amit Pramanik, Ph.D.

ABSTRACT AND BENEFITS

Abstract:
Anaerobic/aerobic (AnA) and completely aerobic (CA) laboratory-scale sequencing batch reactors operating on an acetate- and casamino acids-based synthetic wastewater were used to investigate the suitability of the AnA process for treating nutrient-deficient wastewater in plants that have stringent effluent nutrient requirements. Of particular interest was phosphorus (P)-deficient wastewaters with highly variable influent COD loading that was treated to meet both effluent total suspended solids and P limits.

At a 4-day mean cell residence time, AnA activated sludge had an approximately 20% lower P requirement than CA activated sludge. The difference between the end-of-aerobic cycle polyhydroxyalkanoate and carbohydrate contents of the sludges indicated that the AnA sludge used more influent carbon than the CA sludge for synthesis of non-P-containing storage products. The nitrogen requirements of AnA sludge were similar to those of the CA sludge.

The AnA and CA sequencing batch reactors (SBRs) were subjected to three transient influent COD loading patterns that simulated 1) daily COD loading fluctuations, 2) low weekend COD loading, and 3) extended low COD loading periods. During the loading pattern 1 experiment, the average effluent soluble P concentrations for the AnA and CA SBRs were 0.4 and 1.0 mgP/L respectively, and complete removal of influent acetate was observed. During the loading pattern 2 experiment, the average effluent soluble P concentrations for the AnA and CA SBRs were 0.3 and 0.9 mgP/L respectively, but effluent acetate was detected during the first high COD loading cycle following the low weekend COD loading period. During the loading pattern 3 experiment, the VSS content of both reactors dropped sharply, effluent acetate breakthrough occurred, and effluent P concentrations exceeding 1 mgP/L were detected in both the AnA and CA SBRs. Based on these findings, the AnA process has potential as a technologically and economically superior alternative for wastewater treatment plants treating P-deficient wastewasters to meet stringent effluent TSS and P limits.

Benefits:

- Demonstrates the effect of an initial anaerobic zone on the phosphorus and nitrogen requirements of activated sludge.
- Demonstrates the effect of influent phosphorus and nitrogen concentrations on acetate uptake rates and storage product transformation stoichiometry of anaerobic/aerobic (AnA) and completely aerobic (CA) sequencing batch reactors (SBRs).
- Demonstrates the effect of transient influent COD loading on AnA and CA SBR effluent qualities and sludge storage product content.
- Determined the stability of AnA and CA SBR processes when subjected to long-term variable influent COD loading under phosphorus-limiting conditions.

Keywords: Nutrient deficiency, phosphorus and nitrogen requirements, anaerobic/aerobic activated sludge, polyphosphate-accumulating metabolism

TABLE OF CONTENTS

Acknowledgments iii
Abstract and Benefits iv
List of Tables vii
List of Figures viii
List of Acronyms xii
Executive Summary ES-1

1.0 Background and Rationale 1-1

2.0 Objectives 2-1
2.1 Determine the Effect of an Initial Anaerobic Zone of the Macronutrient Requirements of Activated Sludge 2-1
2.2 Determine the Effect of Limiting N and P Loading on Acetate Uptake Rates and Storage Product Transformation Stoichiometry 2-1
2.3 Determine Whether the Anaerobic/Aerobic Process can more Effectively Treat P-Deficient Wastewater of Variable Influent COD Loading as Compared to the Completely Aerobic Process 2-1

3.0 Results 3-1
3.1 Reactor Operation 3-1
3.1.1. Loading History 3-1
3.1.2. Reactor Solids, Effluent TSS, and Solids Production 3-1
3.1.3. Morphological Observations 3-4
3.1.4. Typical Reactor Profiles 3-4
3.2 Determination of Activated Sludge P Requirements and the Effect of Limiting Influent P Loading 3-6
3.2.1 Effluent Acetate and Soluble COD 3-6
3.2.2 Effluent Soluble Phosphorus 3-6
3.2.3 Acetate Uptake Rates 3-9
3.2.4 Stoichiometry of P Release and Sludge P Content 3-9
3.2.5 Stoichiometry of PHA Formation and CH Degradation 3-9
3.2.6 End-of-aerobic Storage Product Content, Storage Product Yield, and Non-storage Product Yield 3-12
3.2.7 Sludge Volume Index TSS 3-15
3.2.8 Carbon Balance 3-16
3.2.9 COD Balance 3-18
3.2.10 Comparison of Steady-state Profiles with Model Predictions 3-19
3.2.11 Conclusions 3-22
3.3 Determination of Activated Sludge N Requirements and the Effect of Limiting Influent N Loading 3-23
3.3.1 Effluent Acetate and Soluble COD 3-23
3.3.2 Effluent N and Sludge N Content 3-23

3.3.3 Acetate Uptake Rates 3-24
3.3.4 Stoichiometry of P Release 3-24
3.3.5 Stoichiometry of PHA Formation and CH Degradation 3-24
3.3.6 End-of-aerobic Storage Product Content, Storage Product Yield, and Non-storage Product Yield 3-28
3.3.7 SVI and Effluent TSS 3-33
3.3.8 Carbon Balance 3-33
3.3.9 COD Balance 3-33
3.3.10 Conclusions 3-35
3.4 Single-Cycle Transient Loading Experiments and Model Prediction 3-35
3.4.1 Effluent Acetate and Soluble COD 3-36
3.4.2 Effluent Soluble Phosphorus 3-36
3.4.3 Stoichiometry of P Release, PHA Synthesis, and CH Degradation 3-37
3.4.4 End-of-aerobic VSS and Sludge Storage Product Content 3-37
3.4.5 Acetate Uptake Rates 3-37
3.4.6 Conclusions 3-41
3.5 Multiple-cycle Transient Loading Experiments and Model Predictions 3-43
3.5.1 Influent Loading Patterns 3-43
3.5.2 Multiple Cycle Transient Loading Results: Loading Pattern 1 3-44
3.5.3 Multiple Cycle Transient Loading Results: Loading Pattern 2 3-48
3.5.4 Multiple Cycle Transient Loading Results: Loading Pattern 3 3-55
3.5.5 Long-term Simulations 3-63
3.5.6 Conclusions 3-68

4.0 Practical Significance of Research Results 4-1

Appendix A A-1
Appendix B B-1
References R-1

LIST OF TABLES

3-1 Transient Experiment History ... 3-3
3-2 Morphological Observations ... 3-6
3-3 P Mass Balances ... 3-7
3-4 Conditions at P-Deficiency ... 3-15
3-5 N Mass Balances ... 3-24
3-6 Conditions at N-Deficiency ... 3-32
3-7 Daily Average Effluent Soluble P Concentration Loading Pattern 1 Experiment ... 3-45
3-8 Daily Average Effluent Soluble P Concentration Loading Pattern 2 Experiment ... 3-50
3-9 Daily Average Effluent Soluble P Concentration Loading Pattern 3 Experiment ... 3-57

LIST OF FIGURES

3-1a The Multiple Cycle Influent Transient COD Loading Pattern 1 3-2
3-1b The Multiple Cycle Influent Transient COD Loading Pattern 2 3-2
3-1c The Multiple Cycle Influent Transient COD Loading Pattern 3 3-3
3-2 The Typical Experimental Values of Reactor and Effluent Suspended Solids Concentration of an AnA SBR and a CA SBR 3-4
3-3 The Typical Cycle Profile of an AnA SBR 3-5
3-4 The Typical Cycle Profile of a CA SBR 3-5
3-5 The Effect of Influent COD/P Ratio on the Effluent Acetate Concentration of an AnA SBR and a CA SBR 3-7
3-6 The Effect of Influent COD/P Ratio on the Effluent Soluble COD Concentration of an AnA SBR and a CA SBR 3-8
3-7 The Effect of Influent COD/P Ratio on Effluent Soluble P Concentration of an AnA SBR and a CA SBR 3-8
3-8 The Effect of Influent COD/P Ratio on Acetate Removal Rate During the First 110 min of the Cycle of an AnA SBR and a CA SBR 3-10
3-9 The Effect of Influent COD/P Ratio on P release/Acetate-Carbon Removed Ratio and P Content of AnA SBR 3-10
3-10 The Effect of Influent COD/P Ratio on PHA-Carbon Synthesized/Acetate-CarbonRemoved Ratio During the First 110 min of the Cycle of an AnA SBR and a CA SBR 3-11
3-11 The Effect of Influent COD/P Ratio on CH-Carbon Degraded/Acetate-Carbon Removed Ratio During the First 110 min of the Cycle of an AnA SBR and a CA SBR 3-11
3-12 The Effect of Influent COD/P Ratio on End-of-Aerobic Cycle Biomass PHA Content of an AnA SBR and a CA SBR 3-13
3-13 The Effect of Influent COD/P Ratio on End-of-Aerobic Cycle Biomass CH Content of an AnA SBR and a CA SBR 3-13
3-14 The Effect of Influent COD/P Ratio on End-of-Aerobic Cycle Exocellular and Internal Biomass CH Content of an AnA SBR and a CA SBR 3-14
3-15 The Effect of Influent COD/P Ratio on Biomass Storage Product Yield of an AnA SBR and a CA SBR 3-14
3-16 The Effect of Influent COD/P Ratio on Biomass Non-Storage Product Yield of an AnA SBR and a CA SBR 3-16
3-17 The Effect of Influent COD/P Ratio on Sludge Volume Index and Effluent Suspended Solids Concentration of an AnA SBR and a CA SBR 3-17
3-18 The Effect of Influent COD/P Ratio on Carbon Recovery for an AnA SBR and a CA SBR 3-18
3-19 The Effect of Influent COD/P Ratio on COD Recovery for an AnA SBR and a CA SBR 3-19
3-20 The Acetate and P Concentration Profile of AnA SBR: Comparison Between Experimental Data and Model Predictions 3-20
3-21 The PHA and CH Concentration Profile of AnA SBR: Comparison Between Experimental Data and Model Predictions 3-20

3-22 The Acetate and P Concentration Profile of a CA SBR: Comparison Between Experimental Data and Model Predictions 3-21

3-23 PHA and CH Concentration Profile of a CA SBR: Comparison Between Experimental Data and Model Predictions 3-21

3-24 The Time Series PAM and GAM Biomass Fractions Predicted by AnA Model 3-22

3-25 The Effect of Influent COD/N Ratio on Effluent Acetate Concentration from an AnA SBR and a CA SBR 3-25

3-26 The Effect of Influent COD/N Ratio on Effluent Soluble COD Concentration from an AnA SBR and a CA SBR 3-25

3-27 The Effect of Influent COD/N Ratio on Effluent Total Soluble Inorganic N Concentration from an AnA SBR and a CA SBR 3-26

3-28 The Effect of Influent COD/N Ratio on Sludge N Content in an AnA and a CA SBR 3-26

3-29 The Effect of Influent COD/N Ratio on Acetate Removal Rates During the First 110 min of the Cycle of an AnA and a CA SBR 3-27

3-30 The Effect of Influent COD/N Ratio on P Release/Acetate-Carbon Removed Ratio of an AnA SBR 3-27

3-31 The Effect of Influent COD/N ratio on PHA-Carbon Synthesized/Acetate-Carbon Removed Ratio During the First 110 min of the Cycle of an AnA and a CA SBR 3-28

3-32 The Effect of Influent COD/N ratio on CH-Carbon Degraded/Acetate-Carbon Removed Ratio During the First 110 min of the Cycle of an AnA and a CA SBR 3-29

3-33 The Effect of Influent COD/N Ratio on End-of-Aerobic Cycle Biomass PHA Content of an AnA SBR and a CA SBR 3-30

3-34 The Effect of Influent COD/N Ratio on End-of-Aerobic Cycle Biomass CH Content of an AnA SBR and a CA SBR 3-30

3-35 The Effect of Influent COD/N Ratio on EOA Cycle Biomass Exocellular and Internal CH Content of an AnA SBR and a CA SBR 3-31

3-36 The Effect of Influent COD/N Ratio on Biomass Storage Product Yield of an AnA and a CA SBR 3-31

3-37 The Effect of Influent COD/N Ratio on Biomass Non-Storage Product Yield of an AnA and a CA SBR 3-32

3-38 The Effect of Influent COD/N Ratio on Sludge Volume Index and Effluent Suspended Solids Concentration of an AnA SBR and a CA SBR 3-33

3-39 The Effect of Influent COD/N Ratio on Carbon Recovery for an AnA SBR and a CA SBR 3-34

3-40 The Effect of Influent COD/N Ratio on COD Recovery for an AnA SBR and a CA SBR 3-34

3-41 The Effect of Influent COD Peak Loading Factor on Effluent Acetate Concentration of an AnA SBR and a CA SBR 3-36

3-42 The Effect of Influent COD Peak Loading Factor on Effluent Soluble COD Concentration of an AnA SBR and a CA SBR 3-38

3-43 The Effect of Influent COD Peak Loading Factor on Effluent Soluble P Concentration of an AnA SBR and a CA SBR 3-38

3-44 The Effect of Influent COD Peak Loading Factor on P release/Acetate-Carbon Removed Ratio of an AnA SBR 3-39

3-45 The Effect of Influent COD Peak Loading Factor on PHA-Carbon Synthesized Acetate-Carbon Removed Ratio During First 110 Minutes of the Cycle of an AnA SBR and a CA SBR 3-39

3-46 The Effect of Influent COD Peak Loading Factor on CH-Carbon Degraded/ Acetate-Carbon Removed Ratio During First 110 Minutes of the Cycle of an AnA SBR and a CA SBR 3-40
3-47 The Effect of Influent COD Peak Loading Factor on End-of-Aerobic Biomass Concentration of an AnA SBR and a CA SBR 3-40
3-48 The Effect of Influent COD Peak Loading Factor on End-of-Aerobic Biomass PHA Content of an AnA SBR and a CA SBR 3-41
3-49 The Effect of Influent COD Peak Loading Factor on End-of-Aerobic Biomass CH Content of an AnA SBR and a CA SBR 3-42
3-50 The Effect of Influent COD Peak Loading Factor on Acetate Uptake Rates for an AnA SBR and a CA SBR 3-43
3-51 The Effect of Variable Influent COD Loading on Effluent Soluble P Concentration of an AnA SBR and a CA SBR (Loading Pattern 1) 3-44
3-52 The Effect of Variable Influent COD Loading on Effluent Soluble COD Concentration of an AnA SBR and a CA SBR (Loading Pattern 1) 3-46
3-53 The Effect of Variable Influent COD Loading on End-of-Aerobic Biomass Concentration of an AnA SBR and a CA SBR (Loading Pattern 1) 3-46
3-54 The Effect of Variable Influent COD Loading on End-of-Aerobic PHA Content of an AnA SBR and a CA SBR (Loading Pattern 1) 3-47
3-55 The Effect of Variable Influent COD Loading on End-of-Aerobic CH Content of an AnA SBR and a CA SBR (Loading Pattern 1) 3-47
3-56 The Effect of Variable Influent COD Loading on P release from an AnA SBR (Loading Pattern 1) 3-48
3-57 The Effect of Variable Influent COD Loading on Acetate Uptake Rate of an AnA SBR and a CA SBR (Loading Pattern 1) 3-49
3-58 The Effect of Variable Influent COD Loading on Sludge Volume Index and Effluent Suspended Solids Concentration of an AnA SBR and a CA SBR (Loading Pattern 1) 3-49
3-59 The Effect of Variable Influent COD Loading of Effluent Soluble P Concentration of an AnA SBR and a CA SBR (Loading Pattern 2) 3-50
3-60 The Effect of Variable Influent COD Loading of Effluent Acetate Concentration of an AnA SBR and a CA SBR (Loading Pattern 2) 3-51
3-61 The Effect of Variable Influent COD Loading of Effluent Soluble COD Concentration of an AnA SBR and a CA SBR (Loading Pattern 2) 3-52
3-62 The Effect of Variable Influent COD Loading on End-of-Aerobic Biomass Concentration of an AnA SBR and a CA SBR (Loading Pattern 2) 3-53
3-63 The Effect of Variable Influent COD Loading on End-of-Aerobic PHA Content of an AnA SBR and a CA SBR (Loading Pattern 2) 3-53
3-64 The Effect of Variable Influent COD Loading on End-of-Aerobic CH Content of an AnA SBR and a CA SBR (Loading Pattern 2) 3-54
3-65 The Effect of Variable Influent COD Loading on P release from an AnA SBR (Loading Pattern 2) 3-55
3-66 The Effect of Variable Influent COD Loading on Acetate Uptake Rate of an AnA SBR and a CA SBR (Loading Pattern 2) 3-56
3-67 The Effect of Variable Influent COD Loading on Sludge Volume Index and Effluent Suspended Solids Concentration of an AnA SBR and a CA SBR (Loading Pattern 2) 3-56
3-68 The Effect of Variable Influent COD Loading of Effluent Soluble P Concentration of an AnA SBR and a CA SBR (Loading Pattern 3) 3-57
3-69 The Effect of Variable Influent COD Loading of Effluent Acetate Concentration of an AnA SBR and a CA SBR (Loading Pattern 3) 3-58

3-70 The Effect of Variable Influent COD Loading of Effluent Soluble COD Concentration of an AnA SBR and a CA SBR (Loading Pattern 3) 3-59
3-71 The Effect of Variable Influent COD Loading on End-of-Aerobic Biomass Concentration of an AnA SBR and a CA SBR (Loading Pattern 3) 3-60
3-72 The Effect of Variable Influent COD Loading on End-of-Aerobic Biomass PHA Content of an AnA SBR and a CA SBR (Loading Pattern 3) 3-60
3-73 The Effect of Variable Influent COD Loading on End-of-Aerobic Biomass CH Content of an AnA SBR and a CA SBR (Loading Pattern 3) 3-61
3-74 The Effect of Variable Influent COD Loading on P release from an AnA SBR (Loading Pattern 3) 3-61
3-75 The Effect of Variable Influent COD Loading on Acetate Uptake Rate of an AnA SBR and a CA SBR (Loading Pattern 3) 3-62
3-76 The Effect of Variable Influent COD Loading on Sludge Volume Index and Effluent Suspended Solids Concentration of an AnA SBR and a CA SBR (Loading Pattern 3) 3-62
3-77 The AnA Total Biomass Concentration and PAM/GAM Biomass Fraction Profiles 800-Day Simulation (Loading Pattern 1) 3-64
3-78 The AnA End-of-Aerobic Biomass PHA and CH Content Profiles 800-Day Simulation (Loading Pattern 1) 3-64
3-79 The AnA P Release Profile 800-Day Simulation (Loading Pattern 1) 3-65
3-80 The AnA Effluent Soluble P and Acetate Concentration Profiles 800-Day Simulation (Loading Pattern 1) 3-65
3-81 The CA Total Biomass Concentration Profile 800-Day Simulation (Loading Pattern 1) 3-66
3-82 The CA End-of-Aerobic Biomass PHA and CH Content Profiles 800-Day Simulation (Loading Pattern 1) 3-66
3-83 The CA Effluent Soluble P and Acetate Concentration Profiles 800-Day Simulation (Loading Pattern 1) 3-67
3-84 The AnA Total Biomass Concentration and PAM/GAM Biomass Fraction Profiles 800-Day Simulation (Loading Pattern 2) 3-67
3-85 The AnA Effluent Soluble P and Acetate Concentration Profiles 800-Day Simulation (Loading Pattern 2) 3-68
3-86 The AnA P Release and PAM Acetate Uptake Profiles 800-Day Simulation (Loading Pattern 2) 3-69
3-87 The AnA End-of-Aerobic Biomass PHA and CH Content Profiles 800-Day Simulation (Loading Pattern 2)....................... 3-69
3-88 The CA Total Biomass Concentration Profile 800-Day Simulation (Loading Pattern 2) 3-70
3-89 The CA End-of-Aerobic Biomass PHA and CH Content Profiles 800-Day Simulation (Loading Pattern 2) 3-70
3-90 The CA Effluent Soluble P and Acetate Concentration Profiles 800-Day Simulation (Loading Pattern 2) 3-71
3-91 The AnA Total Biomass Concentration Profile 800-Day Simulation (Loading Pattern 3) 3-71
3-92 The AnA Effluent Soluble P and Acetate Concentration Profiles 800-Day Simulation (Loading Pattern 3) 3-72
3-93 The CA Total Biomass Concentration Profile 800-Day Simulation (Loading Pattern 3) 3-72
3-94 The CA Effluent Soluble P and Acetate Concentration Profiles 800-Day Simulation (Loading Pattern 3) 3-73

LIST OF ACRONYMS

AnA	anaerobic/aerobic
CA	completely aerobic
CH	carbohydrate
COD	chemical oxygen demand
d	day
EBPR	enhanced biological phosphorus removal
GAM	glycogen-accumulating metabolism
MCRT	mean cell residence time
N	nitrogen
P	phosphorus
PAM	polyphosphate-accumulating metabolism
PHA	poly-3-hydroxyalkanoate
PHB	poly-3-hydroxybutyrate
PP	polyphosphate
RAS	return activated sludge
SBR	sequencing batch reactor
SPY	storage product yield
SRT	solids retention time
SVI	sludge volume index
TKN	total kjeldahl nitrogen
TSS	total suspended solids
VFA	volatile fatty acid
VSS	volatile suspended solids
WAS	waste-activated sludge

EXECUTIVE SUMMARY

Industrial facilities that generate wastewater must meet secondary treatment standards set forth under the National Pollution Discharge Elimination System (NPDES) permit and monitoring program. Some of these industrial facilites use on-site aerobic biological wastewater treatment processes to reduce the concentrations of soluble and particulate organic pollutants. Many industrial wastewaters are rich in biodegradable organics, and relatively poor in the macronutrients (nitrogen (N) and phosphorus (P) that are essential for biological treatment. N and/or P must be added to biologically treat such nutrient-deficient wastewater.

The variable nature of industrial process operations can subject industrial wastewater treatment plants to transient COD loading conditions (e.g. sudden increases or decreases in influent organic loading). High COD loading events increase the mass of N or P that must be added, while low COD loading events decrease the nutrient demand. Thus, for a wastewater that is both nutrient-deficient and has a highly variable organic loading, the amount of N or P that must be added can vary significantly from time to time. In this type of situation, it is common practice to add N or P in excess so that high organic loadings do not cause nutrient deficiency, but this practice becomes unacceptable when effluent limits on N or P exist, because overaddition of nutrients during low COD loading can result in permit exceedances for N and/or P. Currently, industrial treatment plants with this type of operating problem either operate close to nutrient deficiency or construct a tertiary treatment process downstream of the biological treatment unit where the excess nutrient content is reduced. An example of this is the addition of ferric chloride for P precipitation followed by tertiary filtration. Neither of these two options is attractive. The first is difficult to operate, and risks nutrient deficiency. The second can involve significant capital and operating costs.

For P-deficient wastewaters a third option is proposed that does not suffer from the disadvantages discussed above. In this option, an anaerobic/aerobic activated sludge system would be used. This configuration selects for organisms that accumulate P as polyphosphate (PP) polymers as part of the PP-accumulating metabolism (PAM). In PAM, P release occurs during the initial anaerobic phase, and P uptake occurs during the following aerobic phase. PAM creates an internally-stored reservoir of P (as PP). Under high COD loading conditions, the P released during the anaerobic phase could be available for growth. Under low COD loading, the P is taken up aerobically for growth and PP synthesis. Therefore, for a P-deficient wastewater, P can be added at a steady rate, sufficient to supply P for the average organic loading even though the organic loading is fluctuating. This avoids the difficult operating problem of frequently adjusting the P addition rate to match the influent organic loading.

Under P-limited conditions, the anaerobic/aerobic process has been shown to select for organisms that accumulate carbohydrate (CH) and polyhydroxyalkanoate (PHA), causing a significant fraction of the influent COD to be used for synthesis of organic storage products that do not contain N or P. This storage product accumulation may reduce the relative amount of N or P that must be added to achieve a given amount of COD removal. To assess the suitability of the anaerobic/aerobic process for the treatment of nutrient-deficient wastewater, the effect of the initial anaerobic phase on the macronutrient requirements of activated sludge was explored in this study.

Laboratory-scale sequencing batch reactors (SBRs) were fully automated 1L capacity units operated at sludge age (SRT) of 4d, a hydraulic residence time (HRT) of 12h and with 4 cycles/d. For the anaerobic/aerobic (AnA) SBR each cycle consisted of 30 minute draw and fill, 120 minute anaerobic phase, 180 minute aerobic phase and 30 minute settling. For the fully aerobic (CA) SBR the cycle was the same except that there was no anaerobic phase and the aerobic phase was 300 minutes long. The feed was made up of mineral salts, acetic acid and casamino acids, and various amounts of P (as NaH_2PO_4) and N (as NH_4Cl).

During the P and N requirement determination experiments, the influent COD/P and COD/N ratios were separately varied by adjusting the influent P and N concentrations. At each COD/P and COD/N ratio the SBRs were operated for at least three SRTs (12d) prior to taking measurements. Then measurements were made over a period of 8 d. Influent and effluent were analyzed for soluble P, acetate, soluble COD, TSS (effluent only), NH_3^-N, and NO_3^-N. SBR biomass was analyzed for TSS, VSS, Total P, TKN, polyhydroxyalkanoate (PHA), and carbohydrate (CH). The SBRs were judged to be nutrient deficient when acetate (the major biodegradeable organic component of the influent) could be detected (>5 mg/L) in the effluent.

Experiments were conducted to determine the effect of single-cycle transient COD loading increases (or decreases) on AnA and CA SBRs. In the first set of experiments, the magnitude of the COD loading transient was increased (COD upshift) until effluent acetate was detected. In the second set of transient COD loading experiments, the magnitude of the COD loading transient was decreased (COD downshift) until the COD loading transient was zero (no COD loading for the cycle). To explore the effects of both high and low COD loading, multiple-cycle transient COD loading experiments were conducted in which the daily mass COD loading to the AnA and CA SBRs was added in an unbalanced manner meant to simulate daily COD loading fluctuations.

At a 4 d mean cell residence time, AnA activated sludge had an approx. 20% lower P requirement than CA activated sludge. The difference between the end-of-aerobic cycle PHA and CH contents of the sludges indicated that the AnA sludge used more influent carbon than the CA sludge for synthesis of non-P-containing storage products. The N requirements of AnA sludge were similar to those of the CA sludge, and at the N deficiency point, the storage product contents of both sludges were similar. The AnA and CA SBRs were subjected to three transient influent COD loading patterns that simulated 1) daily COD loading fluctuations, 2) low weekend COD loading, and 3) extended low COD loading periods. During the Loading Pattern 1 experiment, the average effluent soluble P concentrations for the AnA and CA SBRs were 0.4 and 1.0 mg P/L respectively, and complete removal of influent acetate was observed. During the

Loading Pattern 2 experiment, the average effluent soluble P concentrations for the AnA and CA SBRs were 0.3 and 0.9 mg P/L respectively, but effluent acetate was detected after the first high COD loading cycle following the low weekend COD loading period. During the Loading Pattern 3 experiment, the VSS content of both reactors dropped sharply, effluent acetate break-through occurred, and effluent P concentrations exceeding 1 mg P/L were detected in both the AnA and CA SBRs. Long-term model simulations indicated that both SBRs were stable when treating Loading Pattern 1, but that the AnA SBR would gradually fail (due to the presence of elevated soluble P concentrations) after approx. 800 days (the CA SBR biomass was stable, but produced effluent soluble P concentrations of > 1mgP/L throughout the simulation). Based on these findings, the AnA process has potential as a technologically and economically superior alternative for wastewater treatment plants treating P-deficient wastewasters to meet stringent effluent TSS and P limits.

CHAPTER 1.0

BACKGROUND AND RATIONALE

The activated sludge process is often used to treat phosphorus (P)-deficient industrial wastewater. For such applications to be successful, a readily available P source such as phosphoric acid must be added to the wastewater. The amount of P added must be sufficient to provide for cell material synthesis from the influent organic load. Because industrial wastewater organic loads usually are quite variable, it is common practice to add P in excess so that high organic loads do not cause P deficiency, which can lead to sludge bulking.

It is difficult to practice this method of P addition when the plant treating the P-deficient wastewater has discharge criteria for total P. In such plants, it is no longer possible to carry an excess of soluble P in the effluent, because this may lead to violation of the effluent P permit limit. This difficult operating problem exists at treatment plants in industries such as pulp and paper, fruit and vegetable canning, and food and beverage production.

Plants facing this problem have three (expensive or risky) alternatives:

- Equalize influent organic load (expensive)
- Operate on the border between adequate P and P deficiency (risky)
- Add chemicals (i.e., ferric chloride) to precipitate excess P either in a combined or a tertiary treatment process. (expensive)

A fourth unexplored alternative that may be economically and technically superior to these three is:

- Use Enhanced Biological P Removal (EBPR) operating at as low a biomass P content as possible

EBPR plants consist of an initial anaerobic zone that receives wastewater influent and return activated sludge (RAS), followed in sequence by an aerobic zone and a secondary clarifier. In the anaerobic zone, soluble organics are taken up to form polyhydroxyalkaonates (PHAs). This reaction is driven by hydrolysis of microbially-stored inorganic polyphosphates (PP) that produces energy and releases soluble P. Hydrolysis of microbially stored glycogen provides additional energy and reduces power use. In the aerobic basin soluble P is used for biomass growth and stored PP formation. Stored PHAs are used for energy supply, biosynthesis, and glycogen formation.

We propose that when an EBPR system is perturbed by either an increase or a decrease in organic load, the dual use of soluble P (for both PP synthesis and for biomass production) will maintain the secondary effluent soluble P at a low level. Since effluent P permits are almost universally written in terms of total P (including both soluble P and particulate P) it will be important to operate EBPR activated sludge at as a low a P content as possible for the proposed application. Our hypothesis is that it would be appropriate for such an EBPR activated sludge system to be operated with P added to meet the "long-term" average P requirement and to allow the EBPR mechanism to supply P requirements during periods of high organic load fluctuation. We have tested this hypothesis for a range of activated sludge operating conditions and influent loading variations.

Two types of metabolism appear to be important in anaerobic/aerobic (AnA) activated sludge performing EBPR: the PP-accumulating metabolism (PAM; described above) and the glycogen-accumulating metabolism (GAM). The GAM operates in competition with the PAM and functions without the involvement of anaerobic PP hydrolysis and aerobic soluble P uptake. The PAM and GAM are important for the proposed process application. We are forced by effluent total P criteria to operate at low biomass P content and to deal with high influent chemical oxygen demand (COD)/P excursions. These factors will tend to favor significant GAM activity. Thus the competition between these two metabolisms must be carefully considered.

CHAPTER 2.0

Objectives

2.1 Determine the Effect of an Initial Anaerobic Zone on the Nitrogen (N) and P Requirements of Activated Sludge

The macronutrients N and P are required for treatment of nutrient-deficient wastewater, macronutrients. In practice, rule-of-thumb COD/N or COD/P ratios are used to estimate macronutrient requirements, but the actual need depends on cell yield and growth rate. The literature shows that N and P limitations decrease substrate removal rates (Helmers et. al. 1951), but there is no data on the effects of an anaerobic zone on N and P requirements. A goal of this research is to establish the N and P requirements for removal of given mass of COD, and to determine whether the implementation of an anaerobic zone affects these requirements.

2.2 Determine the Effect of Limiting Influent N and P Loading on Acetate Uptake Rates and Storage Product Transformation Stoichiometry

EBPR has been typically studied on wastewater rich in P (and N). Recent research has shown that the influent COD/P ratio affects the anaerobic stoichiometry and kinetics associated with uptake of volatile fatty acids (Lui et. al.1997, Schuler 1998). For application of EBPR to N- or P-deficient wastewaters, the anaerobic-phase EBPR stoichiometry and acetate uptake rates must be established under conditions where the influent N and P contents approach deficient levels. This research will address how PHA and carbohydrate (CH) transformation stoichiometry and acetate uptake rates change as the influent N and P contents approach deficiency for an AnA and completely aerobic (CA) sequencing batch reactors (SBR).

2.3 Determine Whether the AnA Activated Sludge Process Can Treat P-deficient Wastewater of Variable Influent COD Loading more Effectively than the CA Process

The AnA activated sludge process selects for PP-accumulating organisms. Under high COD loading, P that is released during the anaerobic phase should be available for growth during the ensuing aerobic period. Under low COD loading, the P released anaerobically could be used for both growth and PP synthesis, so that low effluent P concentrations are maintained. This study will determine whether the AnA process provides a benefit for treatment of P-deficient wastewater with variable influent COD loading when compared to a CA process. Experimental work will evaluate the effect of transient loading on anaerobic stoichiometry, acetate uptake rates, and storage product levels. Modeling will be conducted to investigate long-term process stability.

CHAPTER 3.0

Results

3.1 Reactor Operation

3.1.1 Loading History

Two laboratory-scale SBRs were operated for 805 days (d) in a range of influent COD loading conditions. Investigators sought to determine the minimum P and N loadings required for complete removal of acetic acid—the primary carbonaceous substrate. The influent COD loading was kept constant during these experiments. The P loading was decreased while N was maintained in excess (influent COD/N ratio approximately 7) from Day 1 through Day 251; the influent N loading was decreased while P was constant and present in excess (influent COD/P ratio approximately 90) from Day 270 through Day 459. At each P and N loading condition, the reactors were operated for at least three MCRTs (12 d) to reach a steady-state condition, and then sampling and analysis of the biomass and wastewater was conducted.

During the period from Day 475 to Day 805, a series of transient COD loading experiments were conducted in which both reactors were subjected to a range of high and low influent COD loadings (Table 3-1). Before each single-cycle transient experiment, the reactor was operated for 3-mean cell residence times (MCRTs) until it reached steady state. Sampling and analysis of the biomass and wastewater was conducted one day prior to the transient test and each day during the transient cycle. Multiple cycle experiments and sampling and analysis of biomass and wastewater were also conducted. Figures 3-1a-c show the transient COD loading patterns tested.

3.1.2 Reactor Solids, Effluent Total Suspended Solids, and Solids Production

Typical reactor and effluent solids values for the AnA and CA SBRs are shown in Figure 3-2. For steady-state conditions with excess P and N, the volatile suspended solids (VSS) content of the AnA and CA SBR were similar: AnA sequencing batch reactor (SBR) averaged 585± 29 (standard deviation [SD]) mg/L; and CA SBR averaged 564±21 mg/L. Effluent total suspended solids (TSS) concentration for both SBRs averaged approximately 20 mg/L. Based on these long-term averages, net solids production for both the AnA and CA SBRs was approximately 0.40 mgVSS/mg COD.

Figure 3-1a. The Multiple Cycle Transient Influent COD Loading Pattern 1.

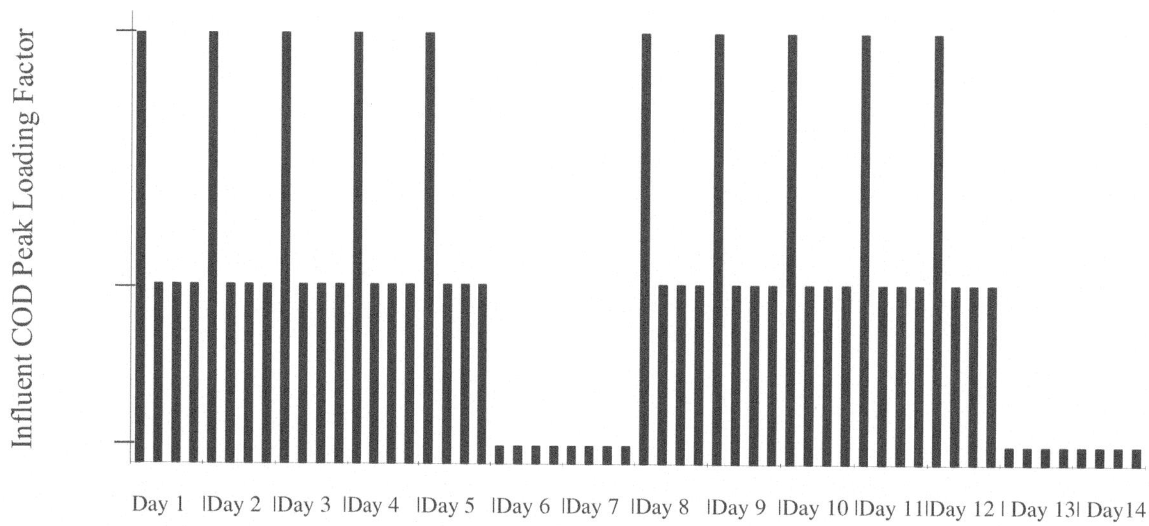

Figure 3-1b. The Multiple Cycle Transient Influent COD Loading Pattern 2.

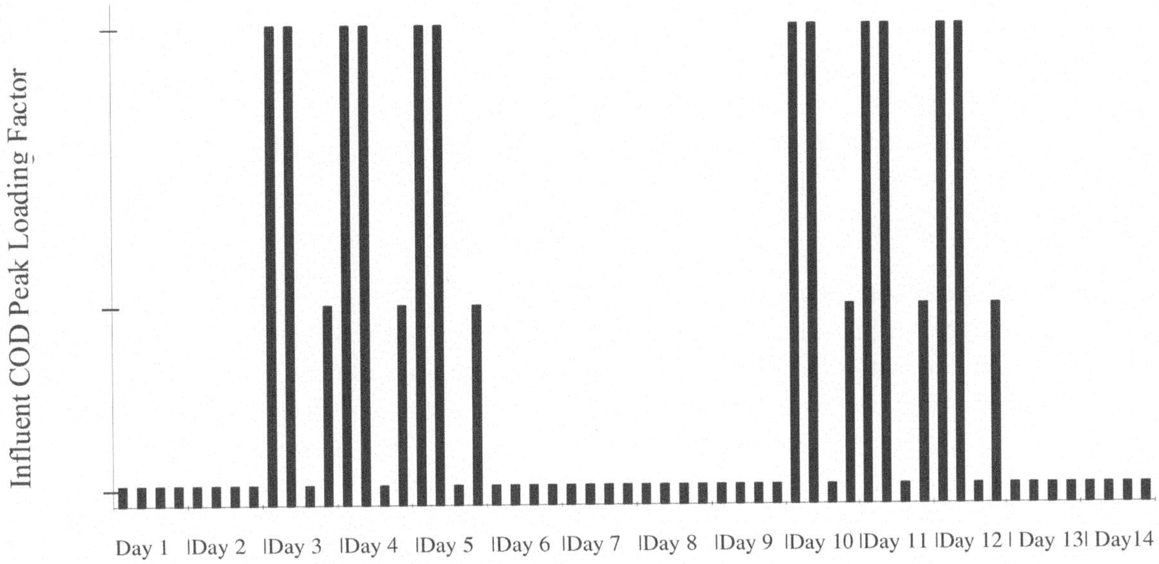

Figure 3-1c. The Multiple Cycle Transient Influent COD Loading Pattern 3.

Table 3-1. Transient Experiment History.

Type of Experiment	Transient Magnitude(s)*	Days of Experiment(s)	Note
Single Cycle, Upshift	1.5x 1.7x 2.0x 2.6x 3.0x 2.3x	475, 492 506, 529 551, 563 579, 603 643, 657 705	Single-Cycle Influent COD Upshift
Single Cycle, Downshift	0.5x 0.0x	667 680	Single-Cycle Influent COD Downshift
Multiple Cycle, Daily COD Loading Fluctuations	1.8x (high) 0.2x (low)	743-756	Loading Pattern 1
Multiple Cycle, Low COD Loading on Weekend	2.4x (high) 0.1x (low)	768-782	Loading Pattern 2
Multiple Cycle, Extended Low COD Loading	2.4x (high) 0.1x (low)	794-805	Loading Pattern 3

* Transient magnitudes shown as ratio of influent cycle COD loading to average cycle COD loading

3.1.3 Morphological Observations

Table 3-2 summarizes morphological observations for the AnA and CA sludges under conditions of excess P and N. Tetrads (clusters of four cells) were observed in both reactors; in the AnA SBR they were present in abundance. The AnA sludge stained Neisser (+) inside the cell, indicating the presence of PP granules, and Gram (-), consistent with previous findings (Mino et. al. 1998). The CA sludge stained Neisser (-), but both Gram (+) and Gram (-) CA organisms were observed. Few filaments were observed in both sludges.

3.1.4 Typical Reactor Profiles

The AnA sludge exhibited the expected EBPR profile (Figure 3-3) showing P release and acetate uptake, coupled with CH degradation and PHA synthesis during the anaerobic phase. During the aerobic phase, P was removed, CH was synthesized, and PHA was degraded. The P content of the AnA sludge will be discussed in section 3.2.4. The CA sludge typically exhibited the profile shown in Figure 3-4. Acetate was completely removed within 170 minutes. During the first 50 minutes of the cycle CH was degraded and PHA was synthesized; during the last 240 minutes of the cycle CH was synthesized, and PHA was degraded. P uptake took place at an approximately constant rate after the first 50 min of the cycle. The P content of the CA sludge was typically 1.5% of VSS.

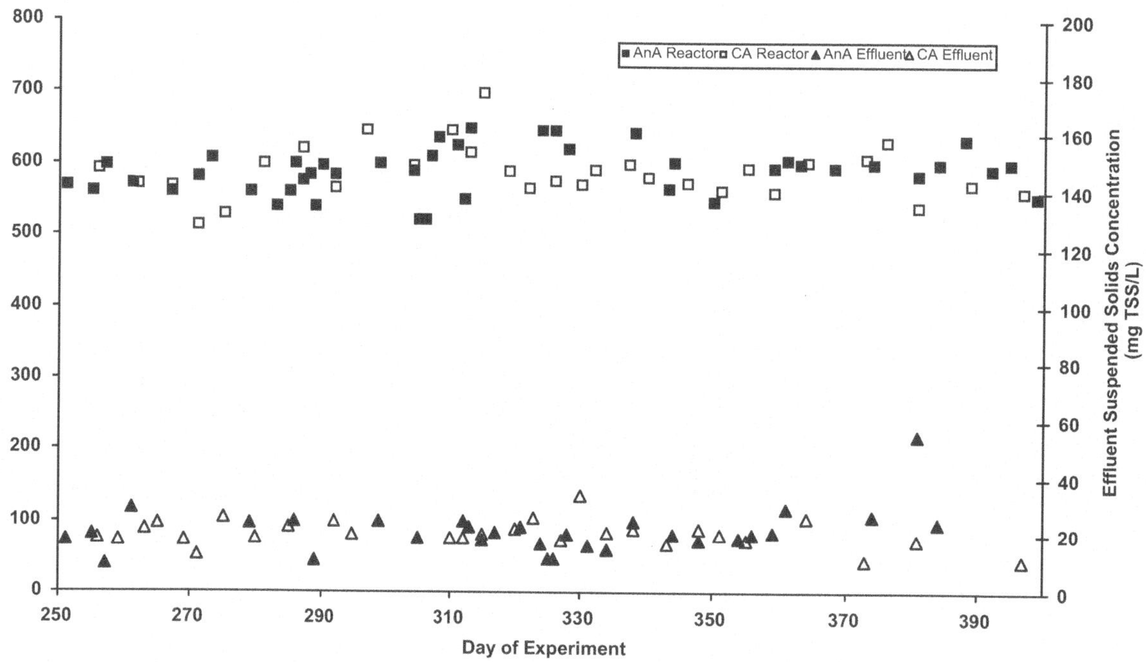

Figure 3-2. The Typical Experimental Values of Reactor and Effluent Suspended Solids Concentration of an AnA SBR and a CA SBR.

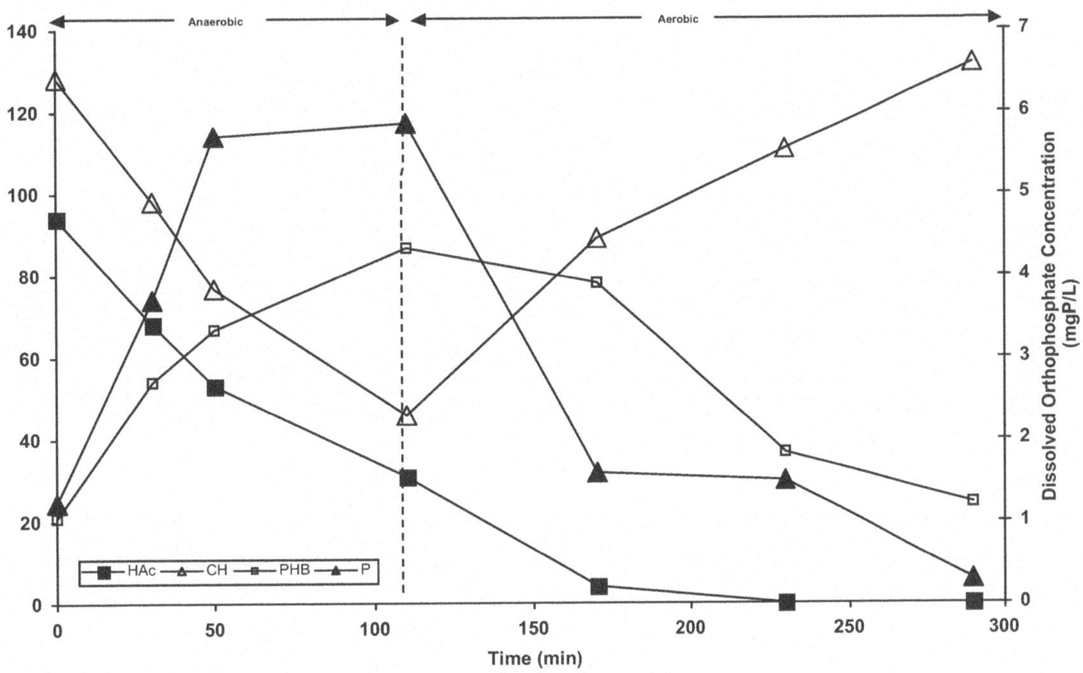

Figure 3-3. The Typical Cycle Profile of an AnA SBR.

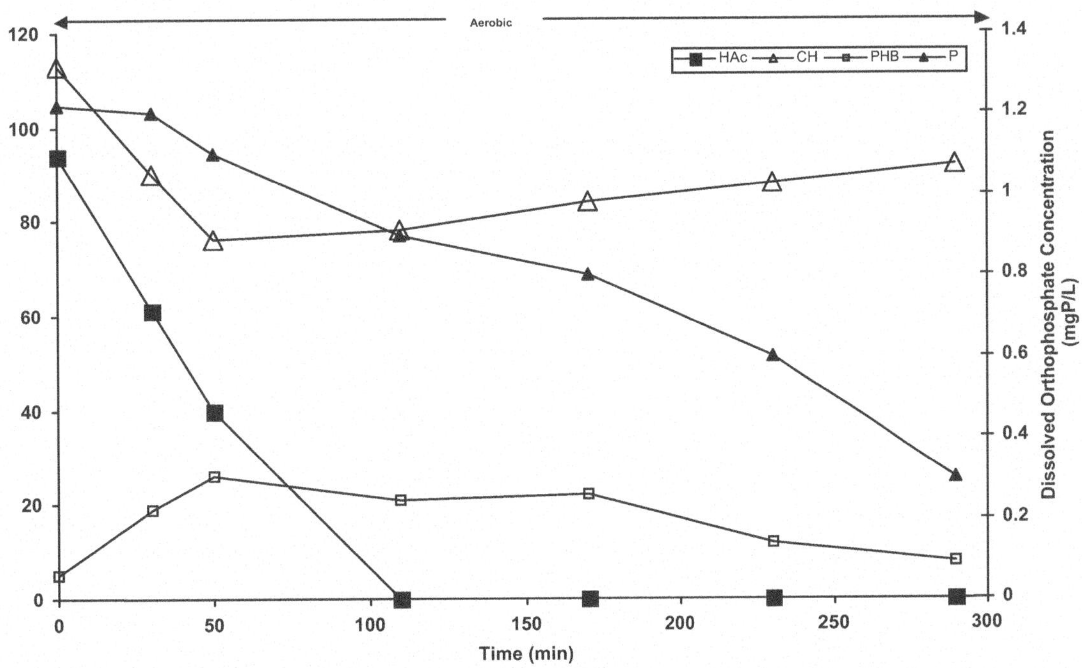

Figure 3-4. The Typical Cycle Profile of a CA SBR.

Table 3-2. Morphological Observations.

Type	AnA	CA
Morphological description	tetrads, few round cocci	long rods, few tetrads
Neisser Stain[1]	+	-
Gram Stain[1]	-	+/-
Filaments	none-few[2]	few[2]
[1]Conducted on sludge taken at the end of the aerobic phase [2]Subjective Scoring of Filament Abundance (Jenkins et al. 1993)		

3.2 Determination of Activated Sludge P Requirements and the Effect of Limiting Influent P Loading

Experiments to determine influent COD/P ratio requirements were conducted under conditions where N was constant and present in excess (influent COD/N ratio was approximately 7). The influent P loading for the AnA and CA SBR was gradually decreased until P deficiency was observed. In practice, low effluent soluble P concentration is used as an indicator of activated sludge P deficiency. For AnA activated sludge, this method is not feasible because the EBPR induced in AnA activated sludge produces low effluent soluble P levels even in the presence of excess P. Therefore, detectable effluent acetate (≥2 mg as acetic acid/L) was used as the P deficiency.

Each influent COD/P ratio was maintained for at least 3 mean cell residence times (MCRTs) in order to achieve steady state prior to taking measurements. At each influent COD/P ratio, mass balances on P were conducted, yielding an average recovery of 94% and 90% for the AnA and CA SBRs, respectively (Table 3-3).

3.2.1 Effluent Acetate and Soluble COD

The AnA SBR was able to operate without exhibiting P deficiency at higher influent COD/P ratios than the CA SBR (Figure 3-5). In the AnA system, acetate did not appear in the effluent until an influent COD/P ratio of 140 was reached; in the CA SBR, effluent acetate was present when the influent COD/P ratio exceeded approximately 110. The effluent soluble COD data agrees with the appearance of acetate in the effluents of both reactors (Figure 3-6). These results show that an AnA SBR can operate at higher influent COD/P ratios than a CA SBR without exhibiting P deficiency, and that the AnA SBR had a P requirement that was approximately 20% lower than that of the CA SBR.

3.2.2 Effluent Soluble P

When the influent COD/P ratio was <100, the effluent soluble P concentration from the AnA SBR was lower than that of the CA SBR (Figure 3-7). This is due to the operation of PAM, which produced the characteristic P release and uptake. Effluent soluble P from the CA approached zero at an influent COD/P ratio of 110, when acetate appeared in the effluent (Figure 3-5). This means that a low effluent P concentration was a good criterion of P deficiency for the CA SBR. The AnA effluent P concentration also approached zero at an influent COD/P ratio of 110.

Table 3-3. P Mass Balances.

Parameter	AnA	CA	AnA	CA	AnA	CA	AnA	CA	AnA	CA	AnA	CA	AnA	CA	AnA	CA
Influent COD/P Ratio	80		90		100		110		120		130		140		160	
Influent P[1]	5.5		4.9		4.4		4.0		3.7		3.4		3.1		2.8	
WAS P[1]	3.5	2.0	3.3	2.1	3.2	2.3	3.0	2.3	2.6	2.2	2.6	-	2.3	-	2.0	-
Effluent P																
Total[1]	1.6	2.9	1.5	2.7	0.9	1.3	0.8	1.1	0.7	0.9	0.7	-	0.6		0.6	
Soluble[1]	0.8	2.6	0.8	2.4	0.2	0.6	0	0.4	0	0	0	-	0.3	-	0.3	-
Recovery,[2] %	93	89	98	98	93	82	95	85	89	84	97	-	94	-	93	-

1 – mg P/d
2 – Recovery = (Influent Total P - Effluent Total P - WAS Total P) x 100 / Influent P

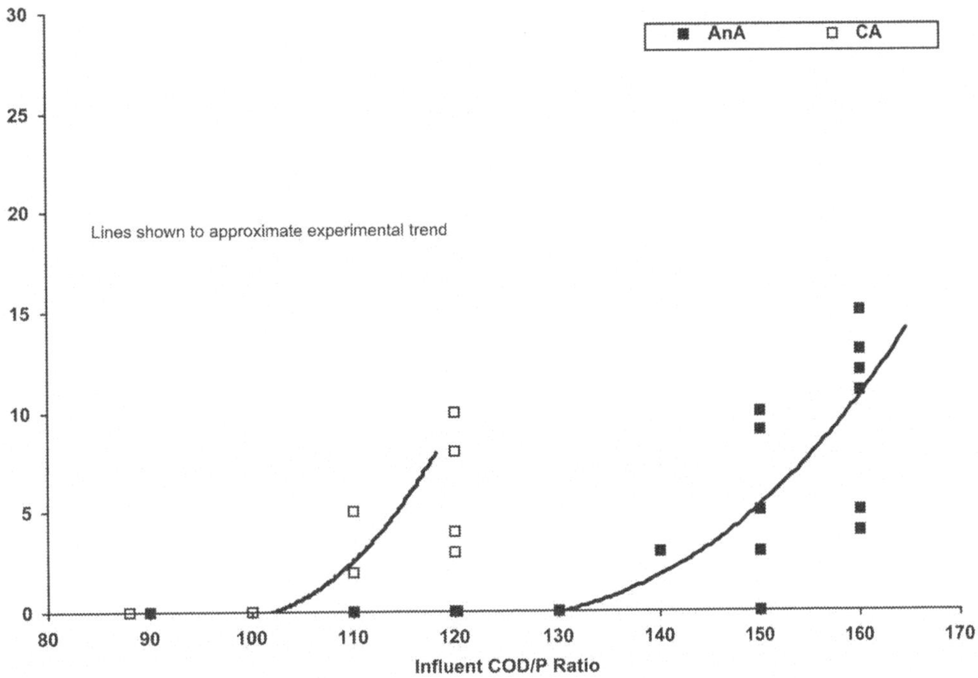

Figure 3-5. The Effect of Influent COD/P Ratio on the Effluent Acetate Concentration of an AnA SBR and a CA SBR.

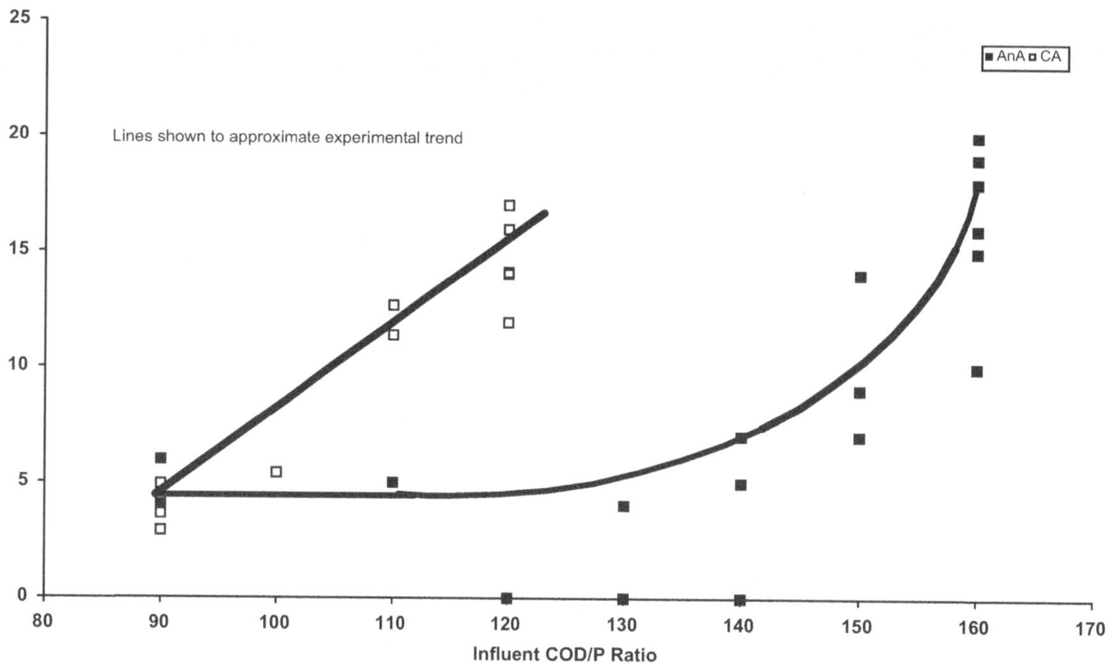

Figure 3-6. The Effect of Influent COD/P Ratio on the Effluent Soluble COD Concentration of an AnA SBR and a CA SBR.

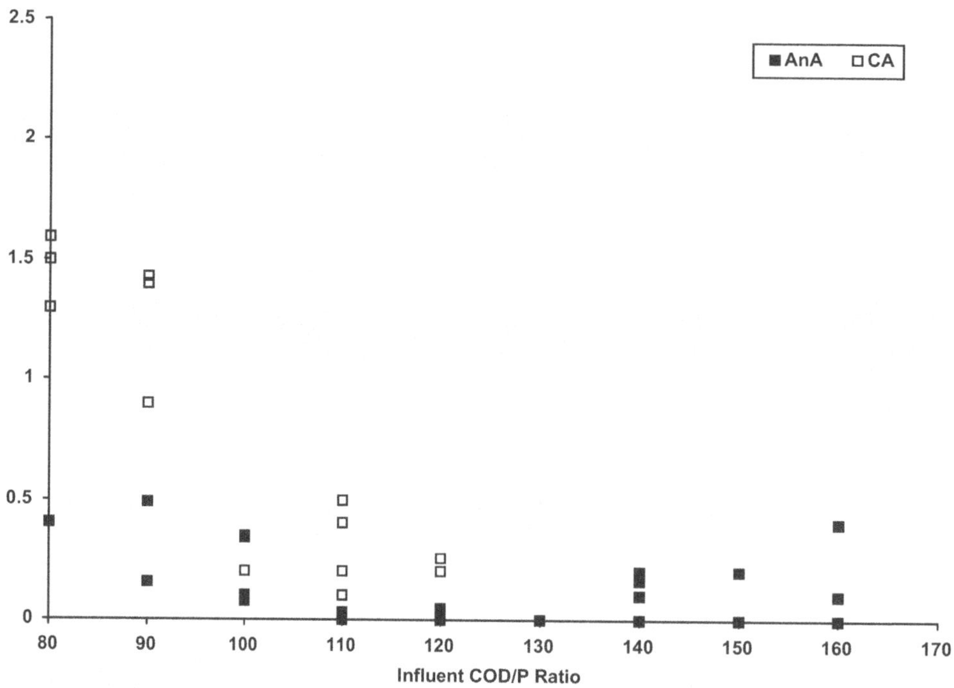

Figure 3-7. The Effect of Influent COD/P Ratio on Effluent Soluble P Concentration of an AnA SBR and a CA SBR.

3.2.3 Acetate Uptake Rates

Acetate removal rates were measured over the first 110 minutes of the cycle, which is the length of the anaerobic phase for the AnA SBR (Figure 3-8). The AnA acetate uptake rate decreased linearly as the influent COD/P ratio increased, which is consistent with previous studies that showed that anaerobic acetate uptake rates decreased as the influent P loading decreased (Lui et. al. 1997, Schuler 1998). The CA acetate uptake rates were initially higher than the AnA SBR when the influent COD/P ratio was ≤ 100, but declined sharply when the influent COD/P ratio was ≥ 110. At P deficiency, the average acetate removal rate was approximately 0.4 mg acetic acid/L/minutes for both sludges.

3.2.4 Stoichiometry of P Release and Sludge P Content

As influent COD/P ratio was increased, the P release/acetate-C removal ratio decreased linearly (Figure 3-9). This was expected, since less P was available for storage as PP as the influent COD/P ratio increased. The sludge P content is shown in Figure 3-9 as biomass P/total end-of-aerobic VSS ratio and as an adjusted P content, or P/total end-of-aerobic non-storage product VSS ratio. The sludge P content (as percent of VSS) ranged between 1% and 1.3%, while the adjusted sludge P content ranged from 1.5% to 2.1%. Thus, P release was observed even though the total P content of the active biomass was close to non-PP containing sludge (Metcalf and Eddy, 1991).

This result, along with the low effluent P results (Figure 3-7) shows that AnA sludge can maintain P release and uptake characteristics up to influent COD/P ratios of 120.

3.2.5 Stoichiometry of PHA Formation and CH Degradation

During the anaerobic phase of the AnA process, acetate-carbon taken into the cell is used to form PHA. The reducing equivalents required for PHA formation are partially derived from the degradation of intracellular CH.

Figure 3-10 shows the effect of influent COD/P ratio on the PHA-C synthesized/acetate-C removed ratio. A higher fraction of influent acetate-carbon is used to form PHA in the AnA SBR than in the CA SBR for all influent COD/P ratios tested. P deficiency had no effect on the poly-3-hydroxybutyrate (PHB)-C synthesized/acetate-C removed ratio for the AnA sludge and PHV-C synthesized/acetate-C removed ratio for both sludges. The CA PHB-C synthesized/acetate-C removed ratio increased from approximately 0.2 to 0.6 mole PHB-C synthesized/mole acetate-C removed as the influent COD/P ratio increased from 80 to 120. Note that the AnA molar ratios of anaerobically synthesized PHB and PHV are consistent with what would be expected of an AnA system that is dominated by the GAM. Figure 3-11 shows the effect of influent COD/P ratio on the CH-C degraded/acetate-C removed ratio for the AnA and CA SBRs. The data shows that CH-C degradation was two to five times higher than that of the CA SBR, that P deficiency had no effect on the CH-C degraded/acetate-C removed ratios, and that the AnA CH degradation stoichiometry was consistent with the operation of a GAM-dominated AnA SBR.

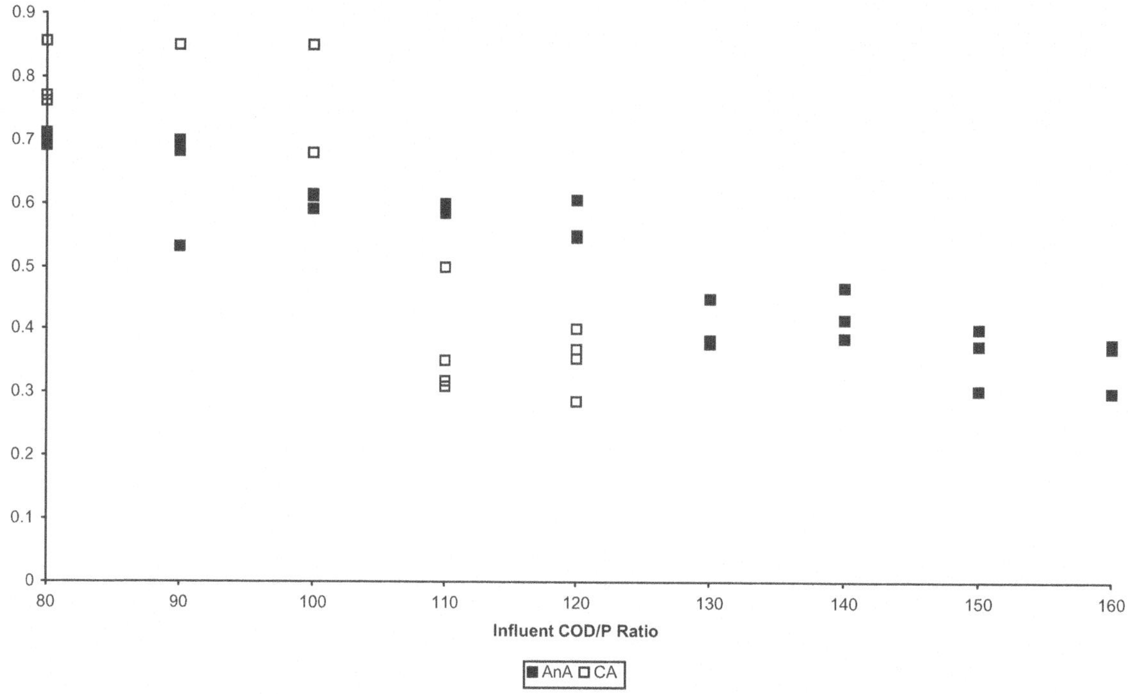

Figure 3-8. The Effect of Influent COD/P Ratio on Acetate Removal Rate During the First 110 Minutes of the Cycle of an AnA SBR and a CA SBR.

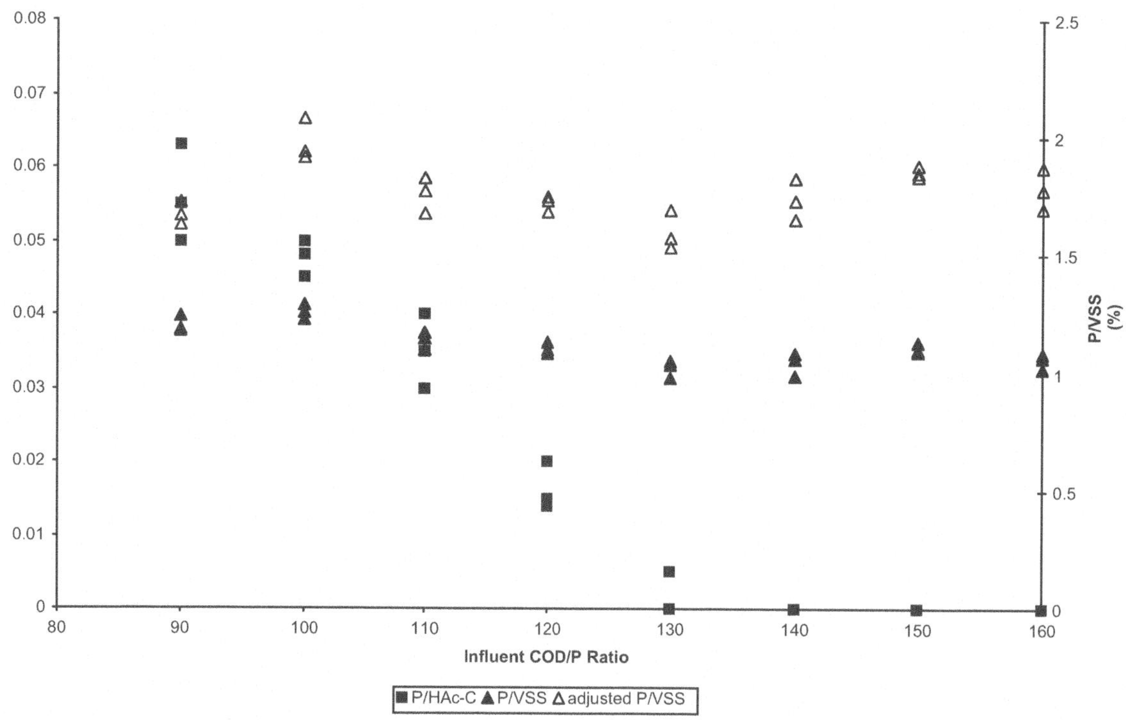

Figure 3-9. The Effect of Influent COD/P Ratio on P release/Acetate-Carbon Removed Ratio and P Content of AnA SBR.

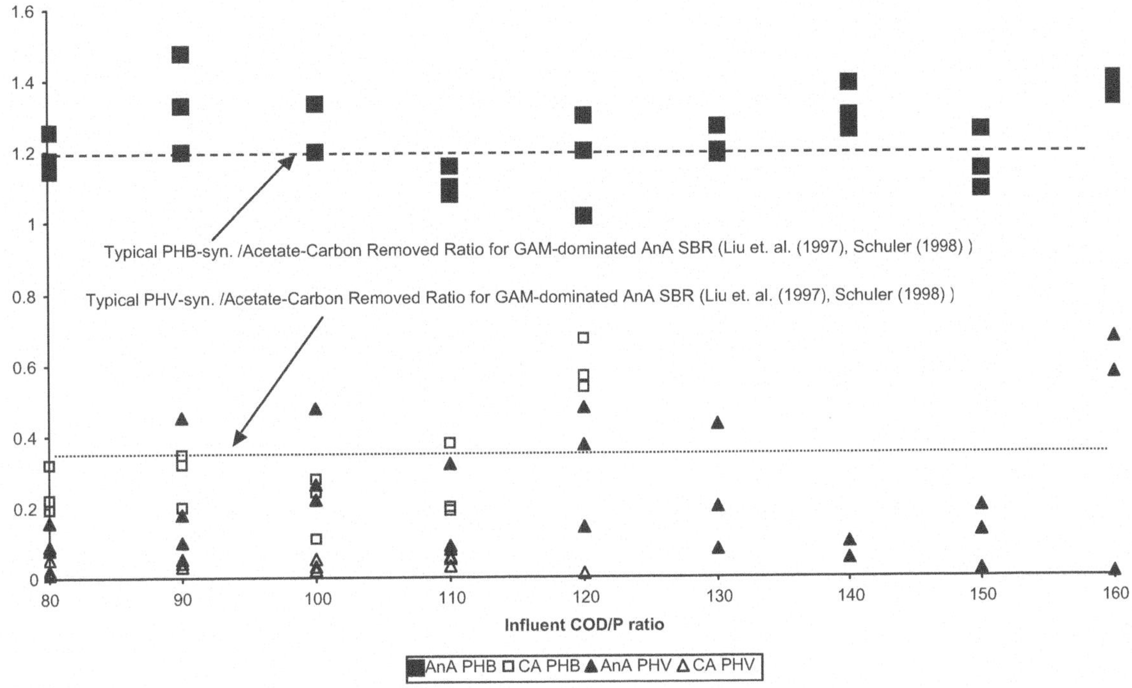

Figure 3-10. The Effect of Influent COD/P Ratio on PHA-Carbon Synthesized/Acetate-Carbon Removed Ratio During the First 110 Minutes of the Cycle of an AnA SBR and a CA SBR.

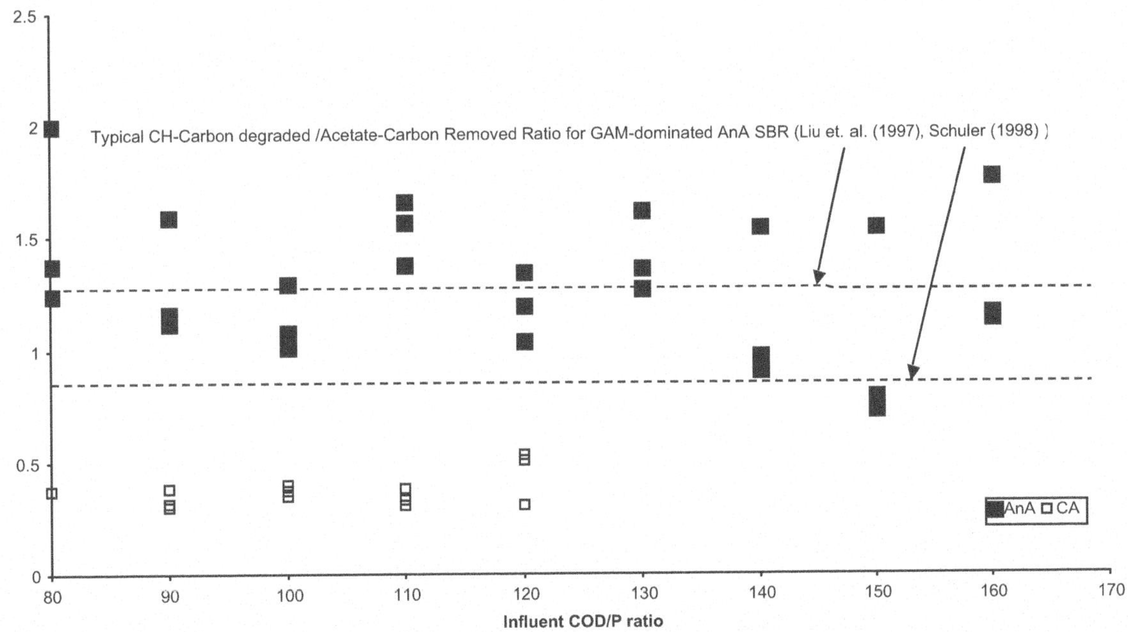

Figure 3-11. The Effect of Influent COD/P Ratio on CH-Carbon Degraded/Acetate-Carbon Removed Ratio During the First 110 Minutes of the Cycle of an AnA SBR and a CA SBR.

3.2.6 End-of-Aerobic Storage Product Content, Storage Product Yield, and Non-Storage Product Yield

The PHA content of the AnA reactor VSS was approximately 4%, while the CA sludge VSS was <1% (Figure 3-12). Increasing the influent COD/P ratio did not significantly affect PHA accumulation for either sludge, but a difference in PHA content of 3–4% was observed for influent COD/P ratios up to 120. The CH content of the AnA sludge increased linearly as the influent COD/P ratio was increased, but that of the CA sludge remained constant (Figure 3-13). This figure shows a biomass CH content difference of approximately 10% for all influent COD/P ratios up to 110 and a difference of approximately 25% when the influent COD/P ratio is ≥ 120. The AnA CH content increased to approximately 35–40% as the influent COD/P ratio was increased above 120.

Biomass-associated CH occurs inside the cell (intracellular CH) and outside the cell as part of bacterial exopolymer (exocellular CH). As the influent COD/P ratio was increased, the exocellular CH content of the AnA sludge increased while the CA sludge remained relatively constant (Figure 3-14). The intracellular AnA and CA CH contents remained relatively constant. The total CH content of the AnA sludge increased as the influent COD/P ratio was increased, but the total CH content of the CA sludge remained constant. These results show that the AnA sludge stored more carbon as both intracellular and exocellular CH than the CA sludge. These results also show that P deficiency caused an increase in the exocellular CH content of the AnA sludge but not the CA sludge.

The storage product data was expressed in terms of storage product yield (spy) using the following equation:

$$spy = [(f_{PHA} + f_{CH})*(f_{VSS})*(C_{eff})*Q_{eff} + (f_{PHA}+f_{CH})(X_{VSS})*Q_{WAS}] / (Q_{in}*C_{COD}) \quad (3\text{-}1)$$

Where:
f_{PHA} = fraction of VSS as total biomass PHA (mg PHA/mg VSS)
f_{CH} = fraction of VSS as biomass CH (mg CH/mg VSS)
f_{VSS} = sludge VSS-to-TSS ratio (mg VSS/mg TSS)
C_{eff} = effluent TSS concentration (mg TSS/L)
Q_{eff} = effluent flow (L/d)
Q_{WAS} = waste activated sludge flow (L/d)
Q_in = influent flow (L/d)
C_{COD} = influent COD concentration (mg COD/L)

The AnA sludge produced more CH and PHA storage product per mg COD removed than the CA sludge for each influent COD/P ratio (Figure 3-15). Given that each sludge was fed the same mass of influent COD per day, these data show that more influent COD was used by the AnA sludge to form organic storage products (CH and PHA). Since neither of these organic polymers requires P for synthesis, this finding is consistent with Figure 3-5, which showed that the AnA SBR required approximately 20% less P than the CA sludge to remove the same amount of acetate. The AnA reactor required less P because more influent COD was used to form organic storage products that do not require P for synthesis.

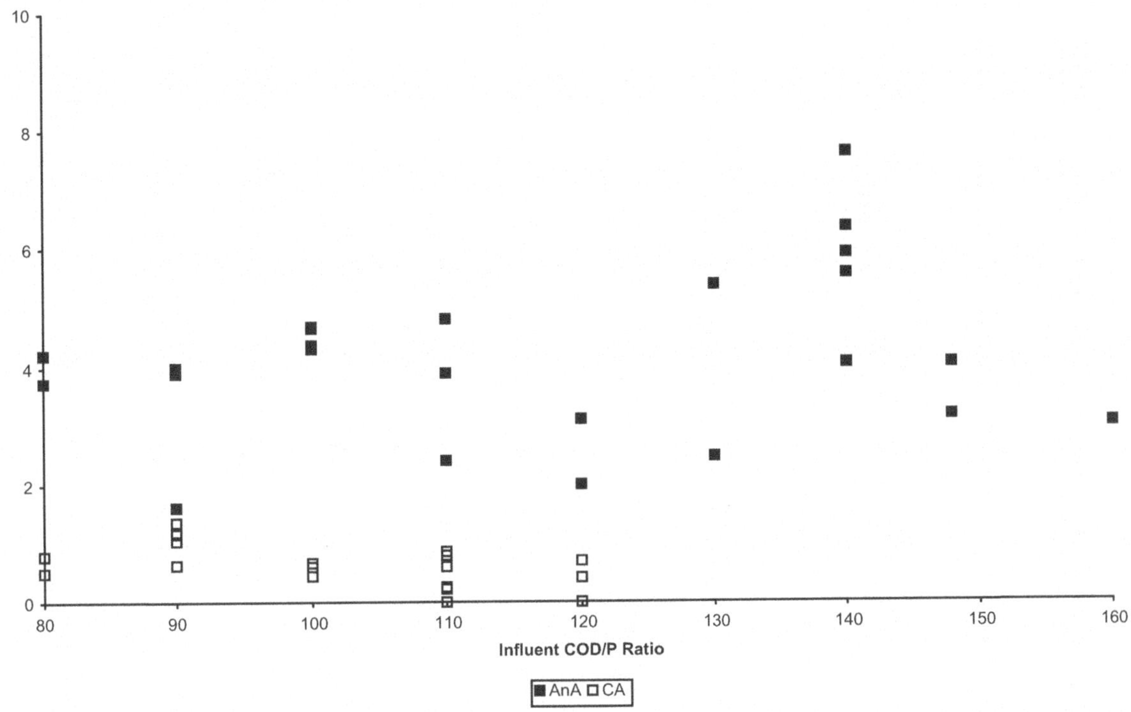

Figure 3-12. The Effect of Influent COD/P Ratio on EOA Biomass PHA Content of an AnA SBR and a CA SBR.

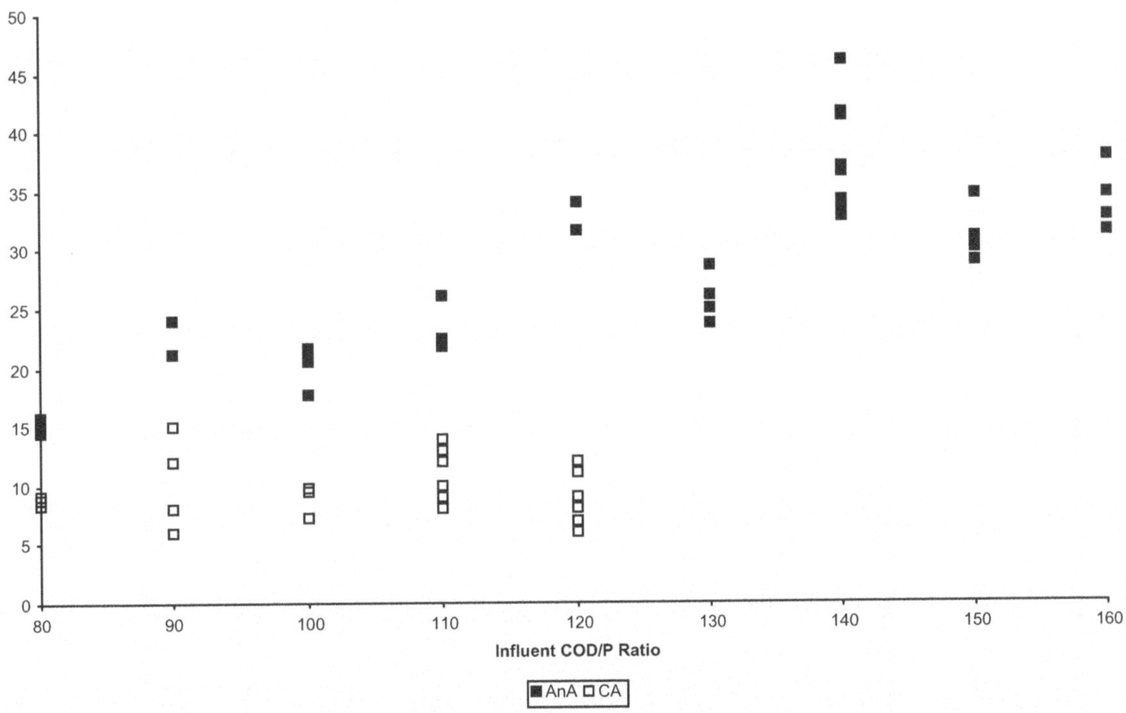

Figure 3-13. The Effect of Influent COD/P Ratio on EOA Biomass CH Content of an AnA SBR and a CA SBR.

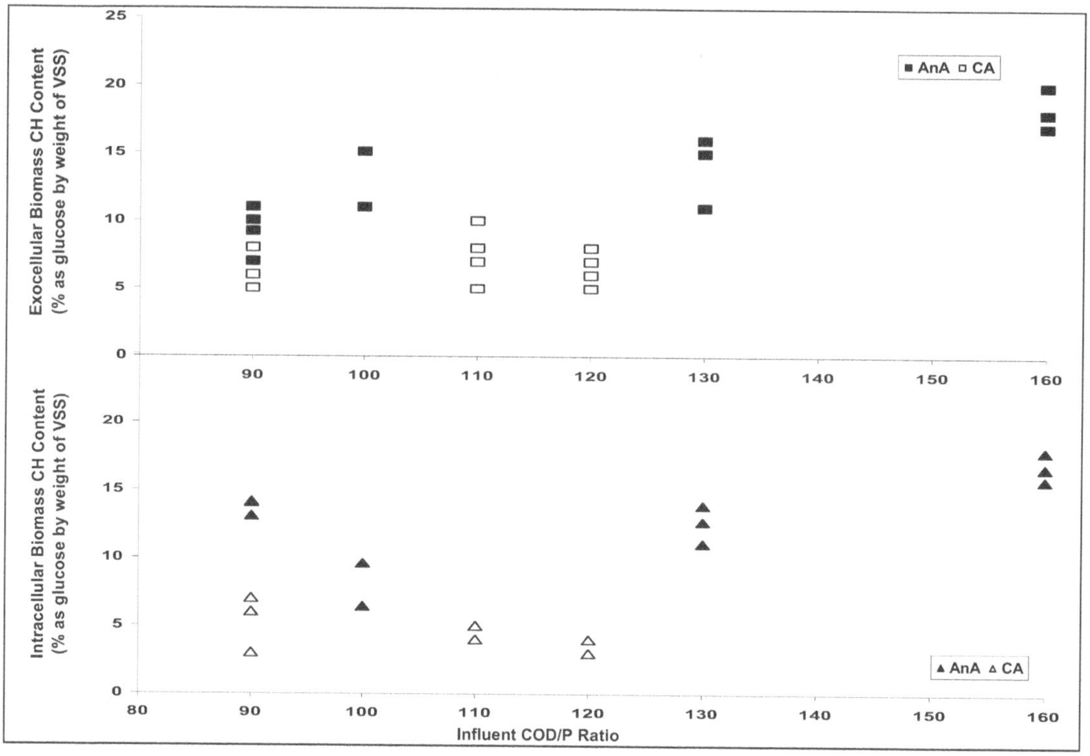

Figure 3-14. The Effect of Infuent COD/P Ratio on Exocellular and Intracellular CH Content of an AnA SBR and a CA SBR.

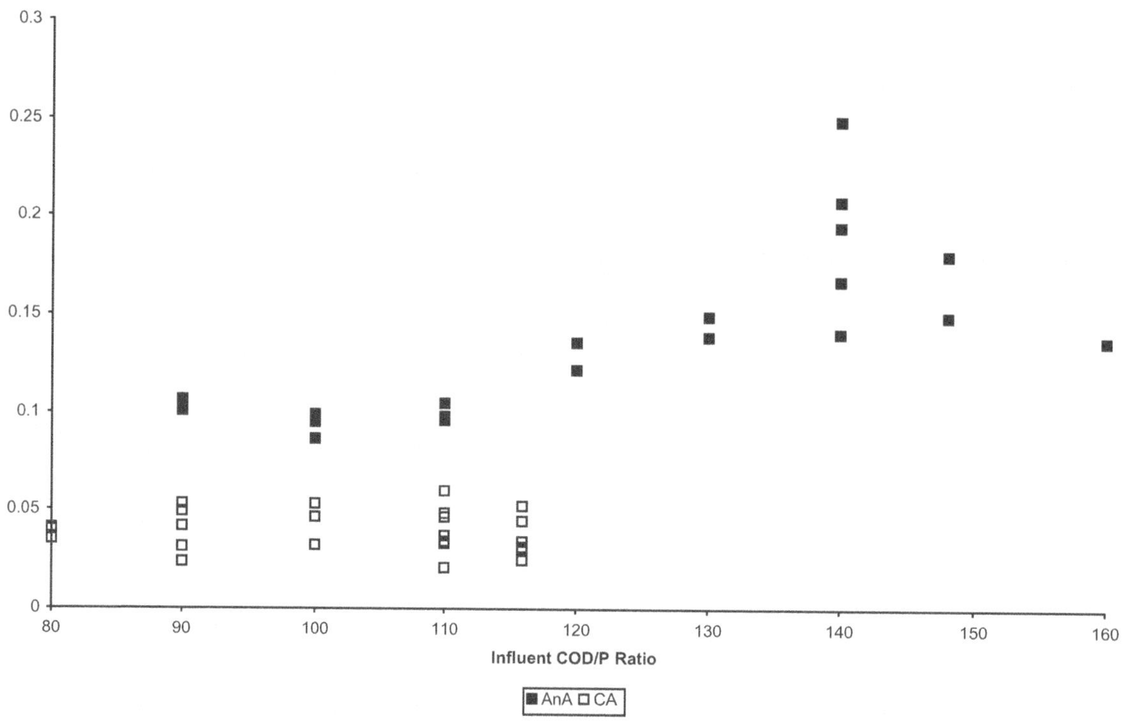

Figure 3-15. The Effect of Influent COD/P Ratio on Biomass Storage Product Yield of an AnA SBR and a CA SBR.

At the influent COD/P ratio where acetate appeared in the effluents (indicating P deficiency), the PHA and CH contents of the CA biomass (VSS basis) were approximately <1% and 15%, respectively, while the PHA and CH contents of AnA sludge were approximately 4% and 35%, respectively. Thus, at the points where the two systems become P-deficient, the differences in their PHA and CH contents were approximately, 4% and 20% respectively. On a COD basis (COD of PHB = 1.38 g/g; COD of CH = 1.07 g/g) these differences are approximately 6% and 22%, respectively. The AnA system biomass used 28% more of the influent COD to form non-P-containing storage products than the CA system biomass. This difference in storage product contents (COD basis) corresponds with the approximately 30:1 ratio observed between the influent COD/P ratios (approximately 140 and 110) at which acetate appeared in the effluents of the AnA and CA SBRs (Table 3-4).

Table 3-4. Conditions at P-Deficiency.

Reactor	PHA (% of VSS)	CH (%VSS)	Storage Product COD (%VSS)	Influent COD/P at P deficiency
AnA	4	35	43	140
CA	<1	15	15	110
Difference	4	20	28	30

While the above data shows that the influent COD/P ratio and the presence of an initial anaerobic zone both influenced the amount of storage products formed, neither of these had an effect on the non-storage product net yield (Figure 3-16). It was constant (between approximately 0.3 and 0.4 mg non-storage product VSS/mg COD removed) over the range of influent COD/P ratios investigated for the CA and AnA SBRs. This data agrees with the theoretical active biomass net yield (0.41 mgVSS/mg COD) of activated sludge growing on acetate with ammonia as the N source (Burkhead and McKinney, 1969).

3.2.7 Sludge Volume Index TSS

The sludge volume index (SVI) values of both SBRs were similar and increased slightly as the influent COD/P ratio increased from 80–120 for the CA sludge and from 80–160 for the AnA sludge (Figure 3-17). Effluent TSS concentrations from both SBRs were similar and generally ranged from approximately 10–30 mg/L (with the exception of one CA outlier). These data indicate that the presence of the anaerobic zone did not provide any benefit with respect to sludge settling or suspended solids removal.

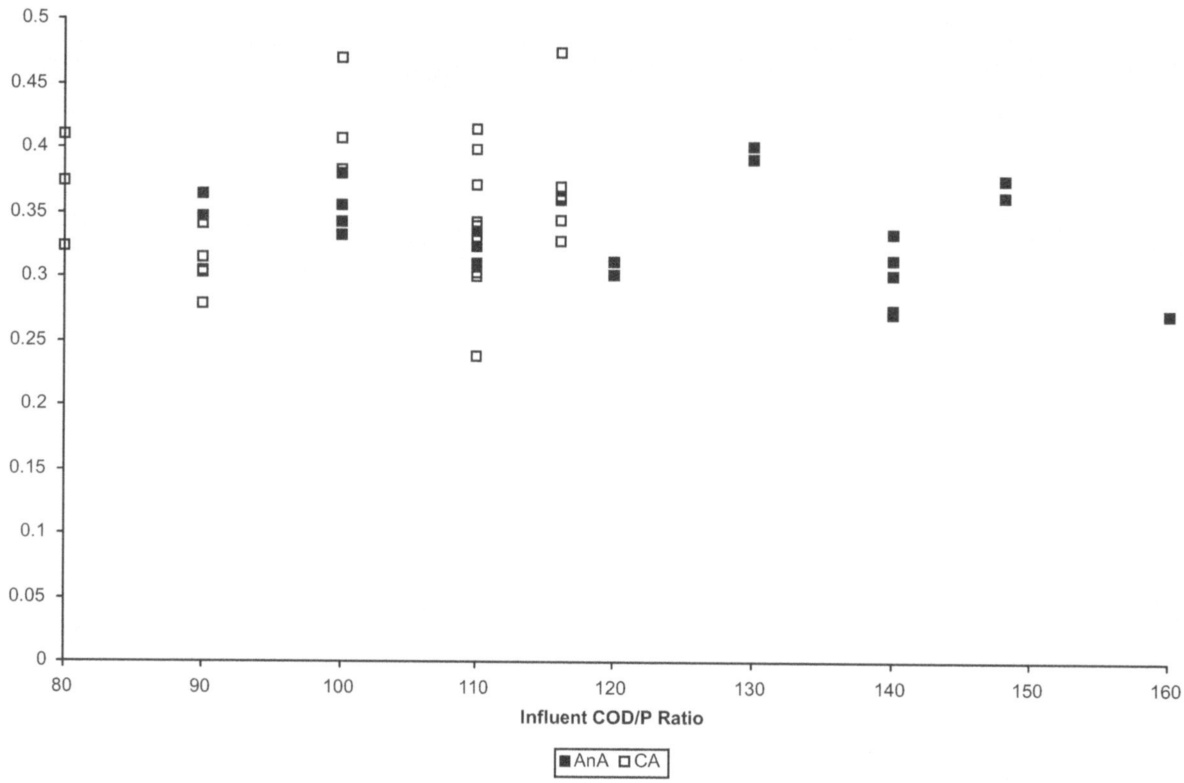

Figure 3-16. The Effect of Influent COD/P Ratio on Biomass Net Non-Storage Product Yield of an AnA SBR and a CA SBR.

3.2.8 Carbon Balance

Carbon balances were performed at each influent COD/P ratio to validate experimental data. Two types of carbon balances were performed. The first was for the first 110 minutes of the cycle; the second for the entire reaction cycle (300 minutes). These carbon balances consisted of balancing the total mass of carbon present at the beginning of the reaction cycle (start) and the total mass of carbon present at the end of the period of interest (e), as expressed in the following:

$$(C_x + C_{CAS} + C_{acetate} + C_{CH} + C_{PHB} + C_{PHV})_{start} = (C_x + C_{CAS} + C_{acetate} + C_{CH} + C_{PHB} + C_{PHV} + C_{CO2})_e \qquad (3\text{-}2)$$

Where:

C_x = active biomass carbon, determined with biomass VSS and storage product data
C_{CAS} = casamino acids carbon, determined with soluble COD data
$C_{acetate}$ = acetate-carbon, determined by direct measurement
C_{CH} = CH biomass carbon, determined by direct measurement
C_{PHB} = PHB biomass carbon, determined by direct measurement
C_{PHV} = PHV biomass carbon, determined by direct measurement
C_{CO2} = carbon dioxide carbon, determined based on stoichiometry

Carbon dioxide production was calculated based of reaction stoichiometry (Schuler 1998) as shown in Appendix A. The carbon recoveries over the anaerobic phase of the AnA SBR were ≥97% (Figure 3-18). This is consistent with previous data showing that AnA SBRs operating at influent COD/P ratios higher than approximately 40 yield carbon recoveries >95%. (Schuler 1998). The carbon recoveries over the first 110 minutes of the CA SBR were approximately 92%. The carbon recoveries over the entire cycle for the AnA SBR were ≥95% until the influent COD/P ratio was 160. Then, the carbon recovery decreased to 85% at the same operating condition where effluent acetate was detected. The reduction in carbon recovery was also seen in the CA SBR. The carbon recovery was 99% when the influent COD/P ratio was ≤100, 95% when the influent COD/P ratio was 110, and 85% when the influent COD/P ratio was 120. The reduction in carbon recovery observed at failure may be due to experimental error.

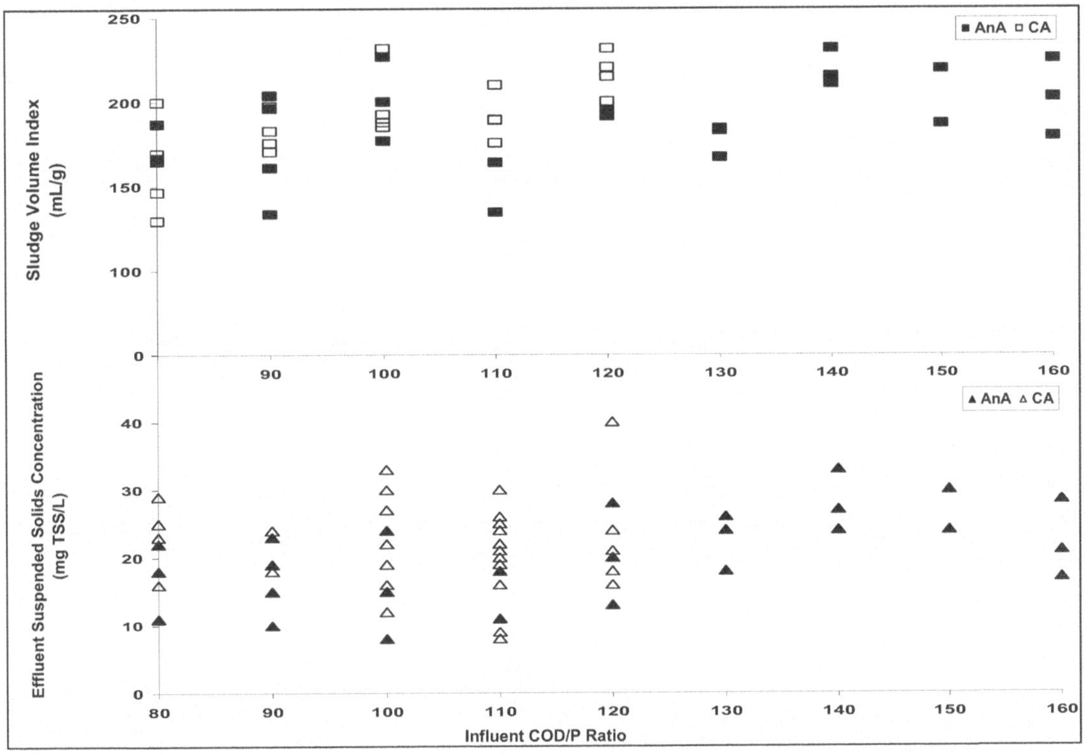

Figure 3-17. The Effect of Influent COD/P Ratio on SVI and Effluent Suspended Solids Concentration of an AnA SBR and a CA SBR.

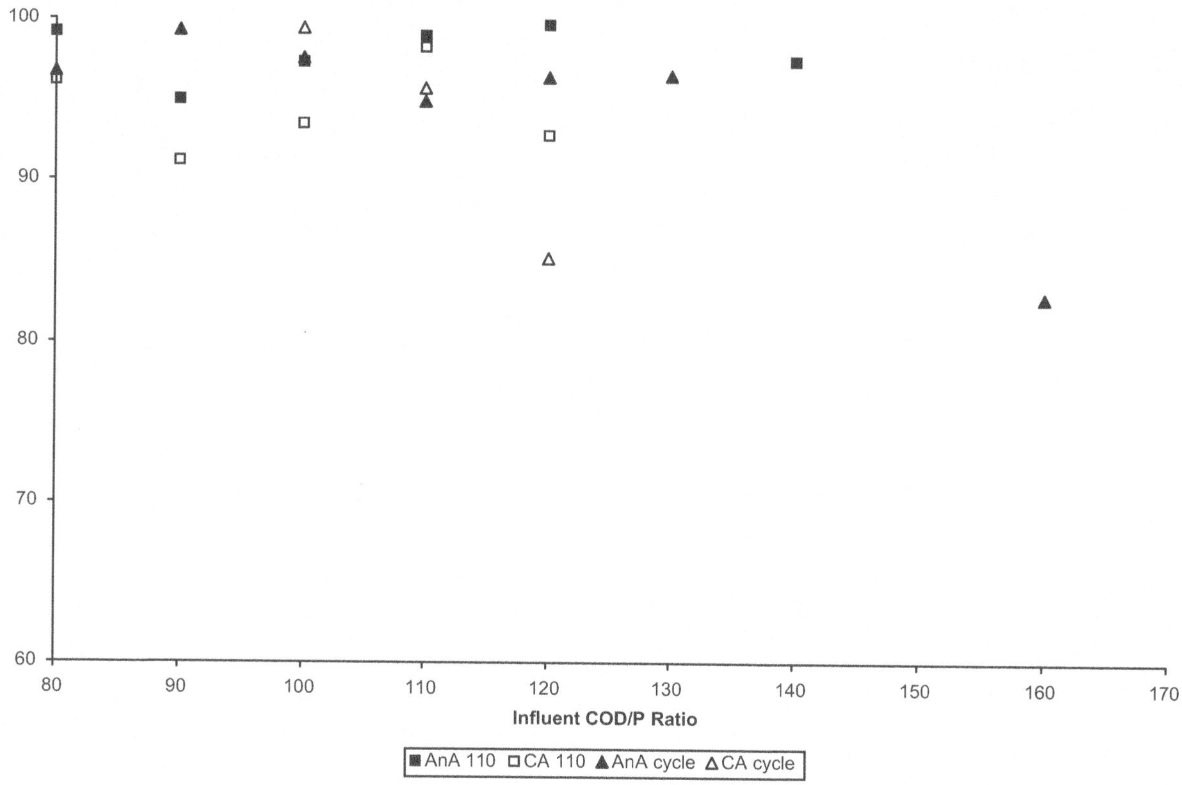

Figure 3-18. The Effect of Influent COD/P Ratio on Carbon Recovery for an AnA SBR and a CA SBR.

3.2.9 COD Balance

COD balances were also performed at each influent COD/P ratio over the first 110 minutes of the cycle and over the entire reaction cycle. The COD balance consisted of balancing the total mass of COD present at the start of the reaction cycle and the total mass of COD present at the end of the period of interest as expressed in the following equation:

$$(COD_x + COD_{CAS} + COD_{acetate} + COD_{CH} + COD_{PHB} + COD_{PHV})_{start} = (COD_x + COD_{CAS} + COD_{acetate} + COD_{CH} + COD_{PHB} + COD_{PHV} + COD_{ox})_{end} \quad (3\text{-}3)$$

Where:

COD_x = COD due to active biomass
COD_{CAS} = COD due to casamino acids
$COD_{acetate}$ = COD due to acetate
COD_{CH} = COD due to biomass CH
COD_{PHB} = COD due to biomass PHB
COD_{PHV} = COD due to biomass PHV
COD_{ox} = COD oxidized during period of interest

COD conversion factors were used to determine the COD of acetate (1.05 mg COD/mg), PHB (1.38 mg COD/mg), PHV (1.6 mg COD/mg) and CH (1.07 mg COD/mg) from direct measurements. The COD due to casamino acids (0.65 mg COD/mg CAS) and active biomass

(1.38 mg COD/mg) were determined by direct measurement. COD oxidation rates were determined by conducting oxygen uptake rate experiments throughout the aerobic phase.

COD recoveries of ≥94% were obtained for both reactors under almost all loading conditions (Figure 3-19). The end-of-aerobic COD recoveries at failure for the CA SBR and AnA SBR were 92% and 88%, respectively.

3.2.10 Comparison of Steady-state Profiles with Model Predictions

Modeling of the AnA and CA SBR was conducted using Matlab 5.3 (Mathworks Inc., Natick, Mass.). The model code consisted of three components: the first specified reactor operating conditions and influent wastewater characteristics; the second solved the non-steady state differential equations associated with removal of soluble constituents and transformation of storage products; and the third performed graphing and reporting functions. The model code, development and calibration are shown in Appendix B.

Good agreement was obtained between model predictions and experimental results for the AnA and CA SBRs at an influent COD/P ratio of 90 (Figures 3-20, 3-21, 3-22, 3-23). The AnA model predicted that the GAM and PAM biomass fractions were approximately 0.85 and 0.15, respectively (Figure 3-24), which is consistent with previous results showing that AnA SBRs operating at influent COD/P ratios >60 are dominated by the GAM (Lui, W.T., 1995; Schuler, 1998).

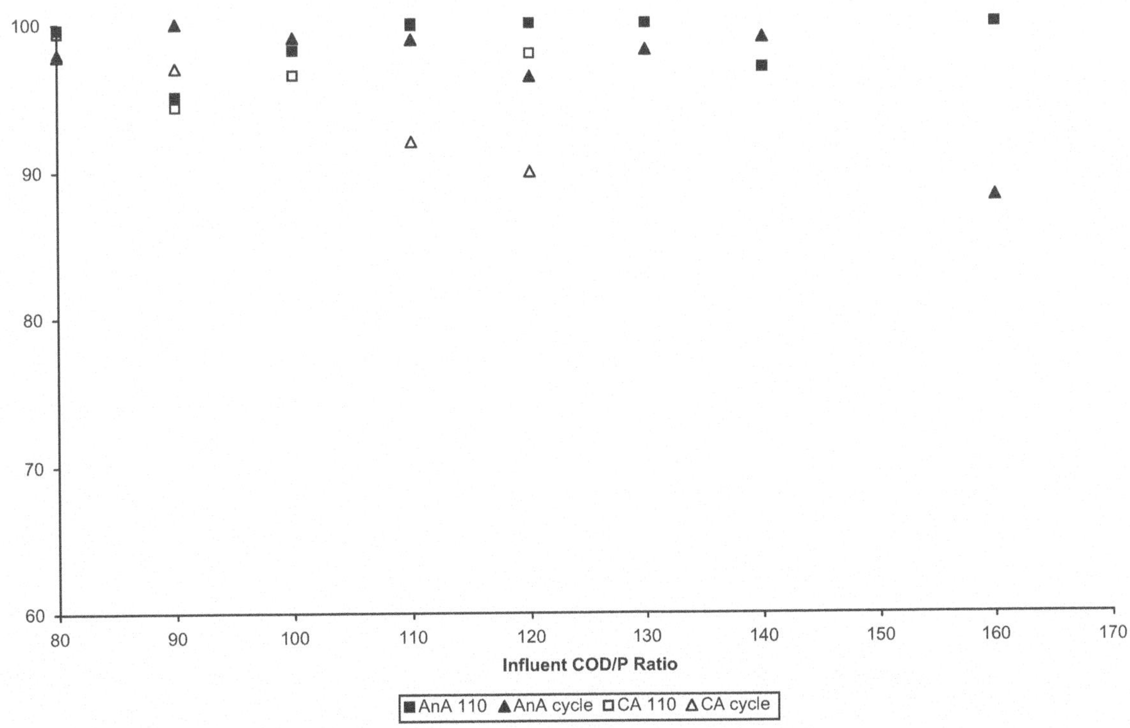

Figure 3-19. The Effect of Influent COD/P Ratio on COD Recovery for an AnA SBR and a CA SBR.

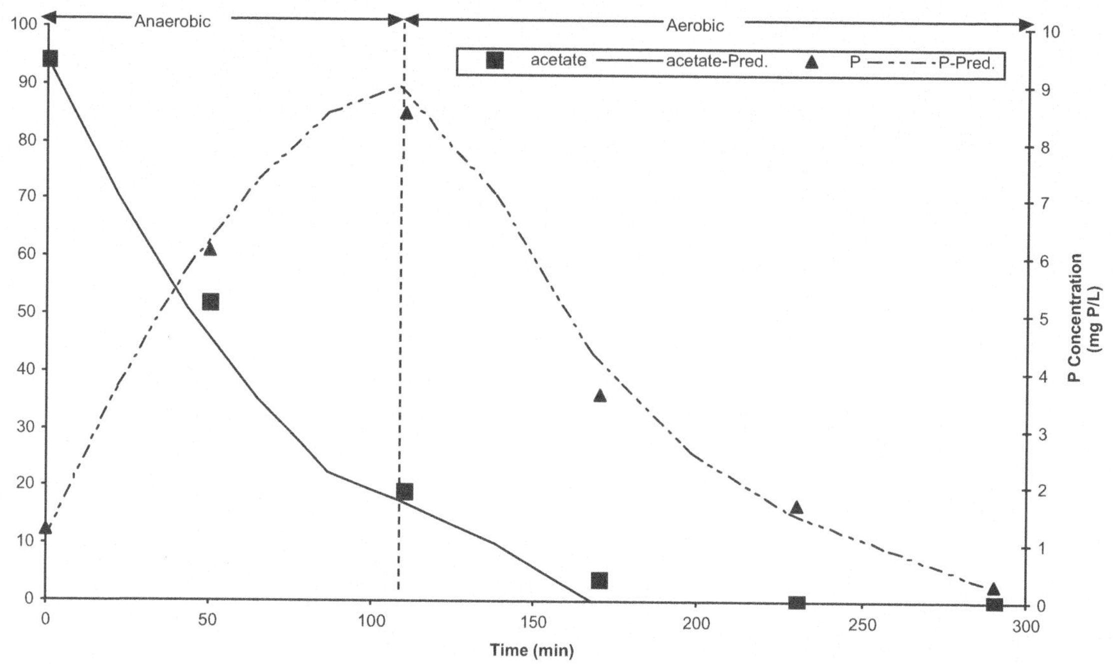

Figure 3-20. The Acetate and P Concentration Profile of AnA SBR: Comparison Between Experimental Data and Model Predictions.

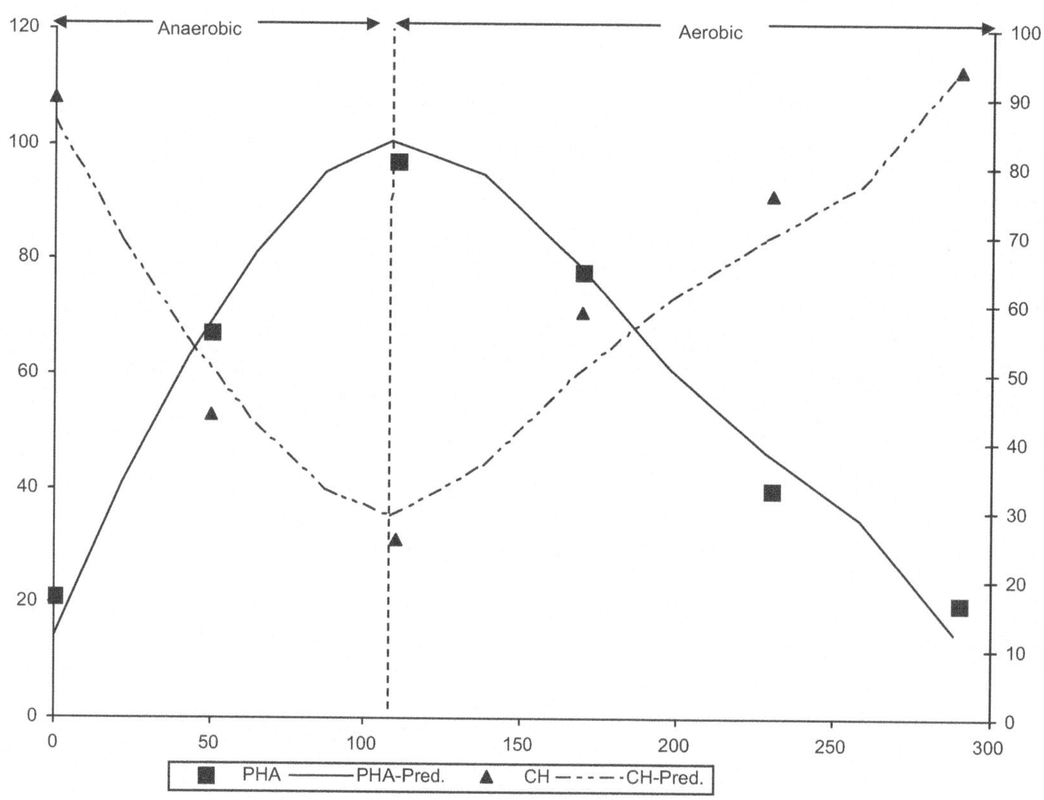

Figure 3-21. The PHA and CH Concentration Profile of AnA SBR: Comparison Between Experimental Data and Model Predictions.

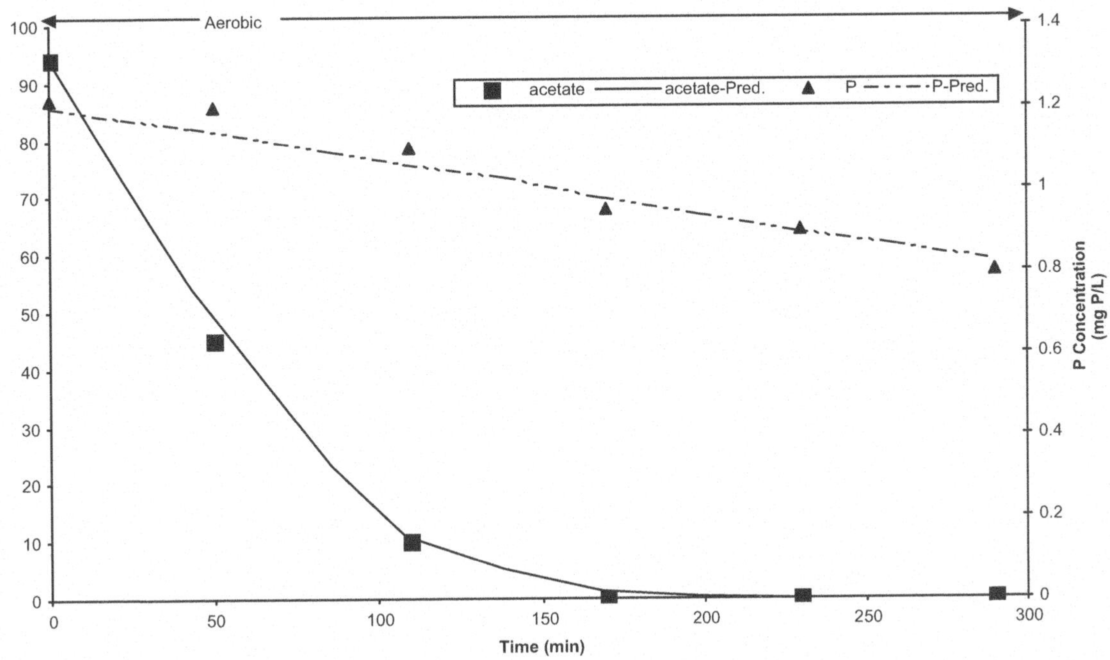

Figure 3-22. The Acetate and P Concentration Profile of a CA SBR: Comparison Between Experimental Data and Model Predictions.

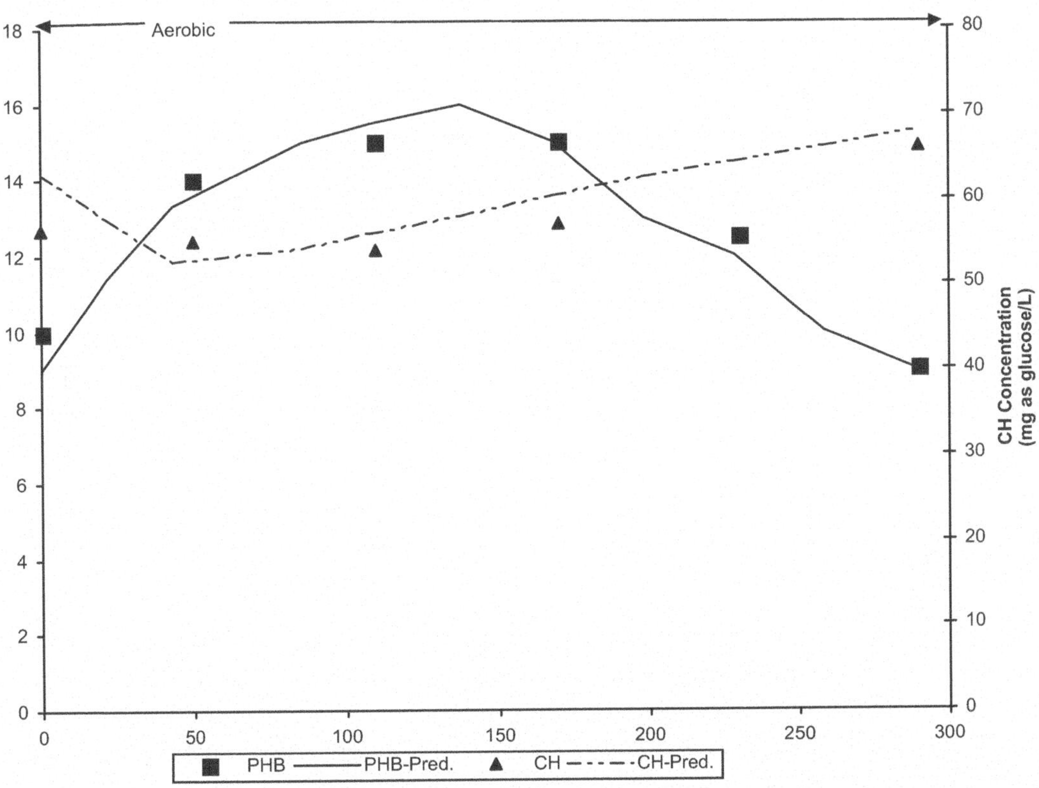

Figure 3-23. The PHA and CH Concentration Profile of a CA SBR: Comparison Between Experimental Data and Model Predictions.

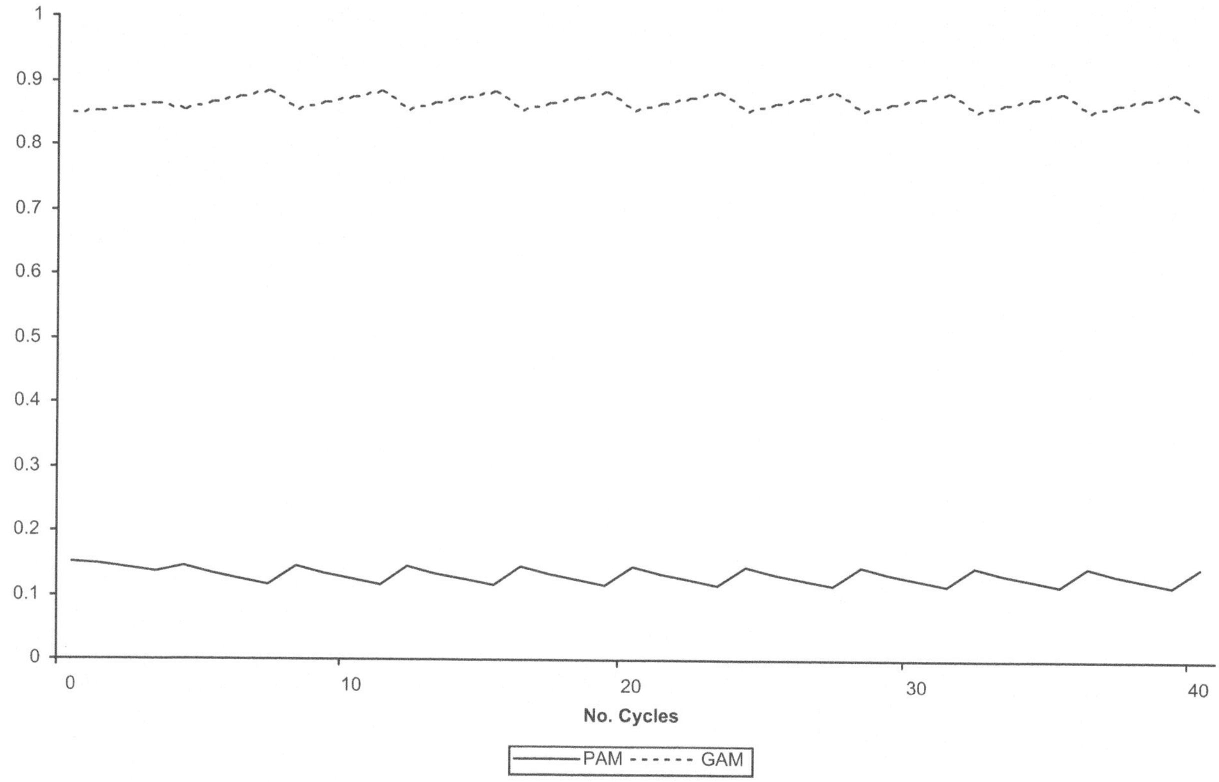

Figure 3-24. The Time Series PAM and GAM Biomass Fractions Predicted by AnA Model.

3.2.11 Conclusions

Measurements of soluble constituents and biomass storage products from an AnA and a CA SBR were validated with mass balances on P, C, and COD. The AnA SBR required approximately 20% less P than the CA SBR. P deficiency occurred at an influent COD/P ratio of 140 for the AnA activated sludge and at an influent COD/P ratio of 110 for the CA activated sludge. The difference in P requirements could be accounted for by the difference in amounts of the non-P containing storage products (PHA and CH). These storage products are important in the metabolism of PAOs and GAOs (Arun et. al. 1988, Satoh et. al. 1992, Cech and Hartman, 1993). PAOs and GAOs are selected in AnA activated sludge, with their relative abundance determined by the influent COD/P ratio of the influent (Liu et. al., 1997; Schuler, 1998). These results suggest that the use of an initial anaerobic zone should in P addition savings for P-deficient wastewaters.

Investigators demonstrated the effect of P deficiency on acetate uptake rates and storage product transformation stoichiometry. The CA acetate uptake rate declined sharply when P deficiency was reached, while the AnA acetate uptake rate gradually decreased as the AnA sludge approached P deficiency. The AnA P release/acetate-C removed ratio linearly declined as the influent COD/P ratio increased, which is consistent with previous results (Lui, 1995; Schuler, 1998). The AnA PHB-C (and PHV-C) synthesized/acetate-C removed ratio was not affected by P deficiency, but the CA PHB-C synthesized/acetate-C removed ratio increased as the influent COD/P ratio increased (the CA end-of-aerobic PHA content did not increase as the influent

COD/P ratio increased, suggesting that nearly all of the PHA synthesized in the CA sludge was degraded prior to the end of the aerobic cycle). The CH-C degraded/acetate-C removed ratio for both sludges was unaffected by P deficiency.

Additional conclusions include the following:

- P release and uptake was observed in an AnA SBR with a sludge P content of between 1% and 1.3% of VSS (adjusted sludge P content ranged from 1.5–2.1% of VSS). Thus, P release can be maintained with low P content sludge.
- The exocellular CH concentration of the AnA sludge increased as the influent COD/P ratio increased, but that of the CA SBR remained constant.
- The SVI and effluent TSS values for both SBRs were similar over the range of influent COD/P ratios tested.
- The stoichiometry of anaerobic P release, PHA synthesis and CH degradation suggest that the GAM was dominant in the AnA SBR.

3.3 Determination of Activated Sludge N Requirements and the Effect of Limiting Influent N Loading

Experiments to determine influent COD/N ratio effects were with P was constant and in excess (influent COD/P ratio approximately 90). The influent N loading for the AnA and CA SBR was gradually decreased until N deficiency was observed (e.g. effluent acetate $\geq$ 2 mg as acetic acid/L). Each influent COD/N ratio was maintained for at least 3 MCRTs to achieve steady state prior to making measurements. Mass balances for N at each influent COD/N ratio gave an average recovery of 91% and 89% for the AnA and CA SBRs respectively (Table 3-5).

3.3.1 Effluent Acetate and Soluble COD

Acetate appeared in both the AnA and CA SBR effluents at influent COD/N ratios of $\geq$ 30 (Figure 3-25). The effluent soluble COD data was consistent with effluent acetate data (Figure 3-26). This result shows that N deficiency occurred at the same influent COD/N ratio for both SBRs, and therefore, the N requirement for the AnA and CA SBR was the same.

3.3.2 Effluent N and Sludge N content

By the common criterion for N deficiency (effluent (NH_3-N + NO_3-N) <1.0 mg N /L), both the CA and AnA SBRs were N-deficient at an influent COD/N ratio of 20 (Figure 3-27). Thus, the common index for N deficiency occurs at a lower influent COD/N ratio (20) than the N deficiency criterion based on appearance of effluent acetate (from Figure 3-25). Using this observation, the influent COD/N ratio range was divided into three regions. Region I was below an influent COD/N ratio of 20 and was the "Excess N Region." Region II was between influent COD/N ratios of 20 and 30 and it was the "N-Limited Region." The region where the influent COD/N ratio was >30 (Region III) was the "N-Deficient Region."

Figure 3-28 shows the total sludge N content (as percent of sludge VSS) and the adjusted sludge N content (as percent of non-storage product VSS) for the AnA and CA sludges. The adjusted N content of both sludges gradually increased in the N-limited and N-deficient regions, ranging from 9–12% of VSS, which is consistent with typical bacterial cell N content (Lehninger 1993).

The total sludge N content of the VSS in both reactors changed under N-limited conditions. It was approximately 8% for both sludges in Region I, but in Region II, the N content of both sludges decreased to approximately 6.5%. This change in sludge N content is attributable to changes in storage product content (see Section 3.3.6.).

3.3.3 Acetate Uptake Rates

The AnA and CA acetate uptake rates were constant throughout Regions I and II (Figure 3-29). When the influent COD/N ratio was 30, the acetate uptake rate for both sludges decreased by approximately 50%—an observation consistent with the appearance of acetate in the effluent. Thus, N deficiency but not N limitation greatly reduced the acetate uptake rates of both sludges, a result previously found by others (Sawyer, 1941; Helmers et. al., 1951; Jones, 1964).

3.3.4 Stoichiometry of P Release

The anaerobic P release/acetate-C removed ratio was relatively constant in Regions I and II, but decreased slightly under N-deficient conditions (Region III, Figure 3-30).

3.3.5 Stoichiometry of PHA Formation and CH Degradation

Figure 3-31 shows the effect of influent COD/N ratio on the PHA-C synthesized/acetate-C removed ratio. The data shows that the fraction of influent acetate-C used to form PHA in the AnA and CA SBRs depends on the operating region. In Region I, more PHA-C synthesis/acetate-C removed occurred in the AnA SBR, consistent with the results of the P requirement experiments (Section 3.2.5.). Under N-limiting conditions (Region II),

Table 3-5. N Mass Balances.

Parameter	AnA	CA	AnA	CA	AnA	CA	AnA	CA	AnA	CA	AnA	CA	AnA	CA	AnA	CA
Influent COD/N Ratio	6.9		16		20		23		27		30		33		36	
Influent TKN[1]	63.8		28.0		22		19		16.4		14.6		13.2		12.2	
WAS TKN[1]	16.2	13.7	12.4	12.4	12.5	9	10.5	9.4	9.4	13.8	11.8	11.5	12.5	11.2	9	11.3
Effluent TKN																
Total[1]	41.3	9.0	3.8	3.8	2.9	2.7	2.2	2.3	2.3	4.9	4.0	4.0	2.9	3.7	2.7	2.7
Soluble[1]	38	4.7	0.6	0.6	0.6	0.4	0.4	0.3	0.3	1.2	1.0	1.1	0.6	0.4	0.4	0.2
Effluent NO_3-N[1]	1.0	0.7	0.7	0.7	0	0.7	0	0	0	1.0	1.3	1.3	0	0	0.7	0
Recovery (%[2])	93	83	76	76	94	87	96	97	97	89	90	88	94	91	87	96

1 – mg N/d
2 – % Recovery = (Influent TKN – Effluent Total TKN – Effluent NO_3-N – WAS TKN) x 100/ Influent TKN
Recoveries at Influent COD/N Ratios of 12 and 18 were 95% and 87% (AnA), 84 and 94% (CA) (data not shown due to space constraints)

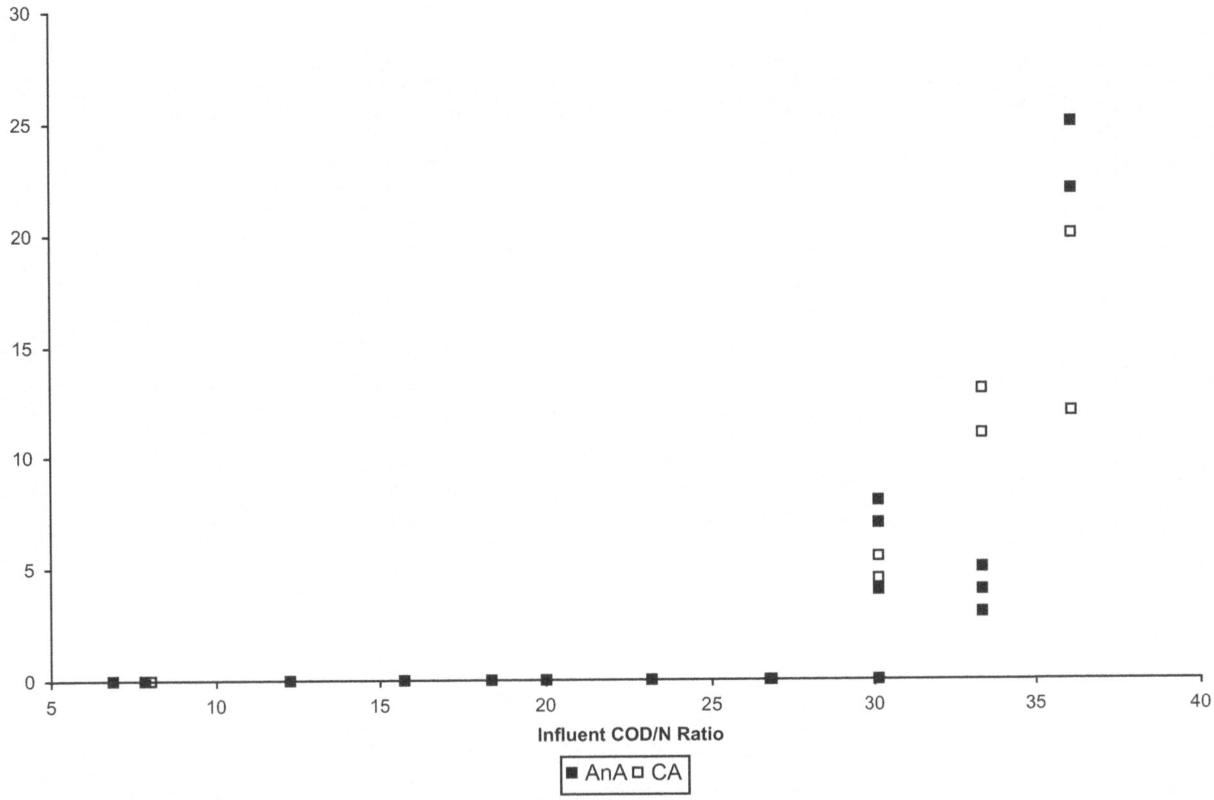

Figure 3-25. The Effect of Influent COD/N Ratio on Effluent Acetate Concentration from an AnA SBR and a CA SBR.

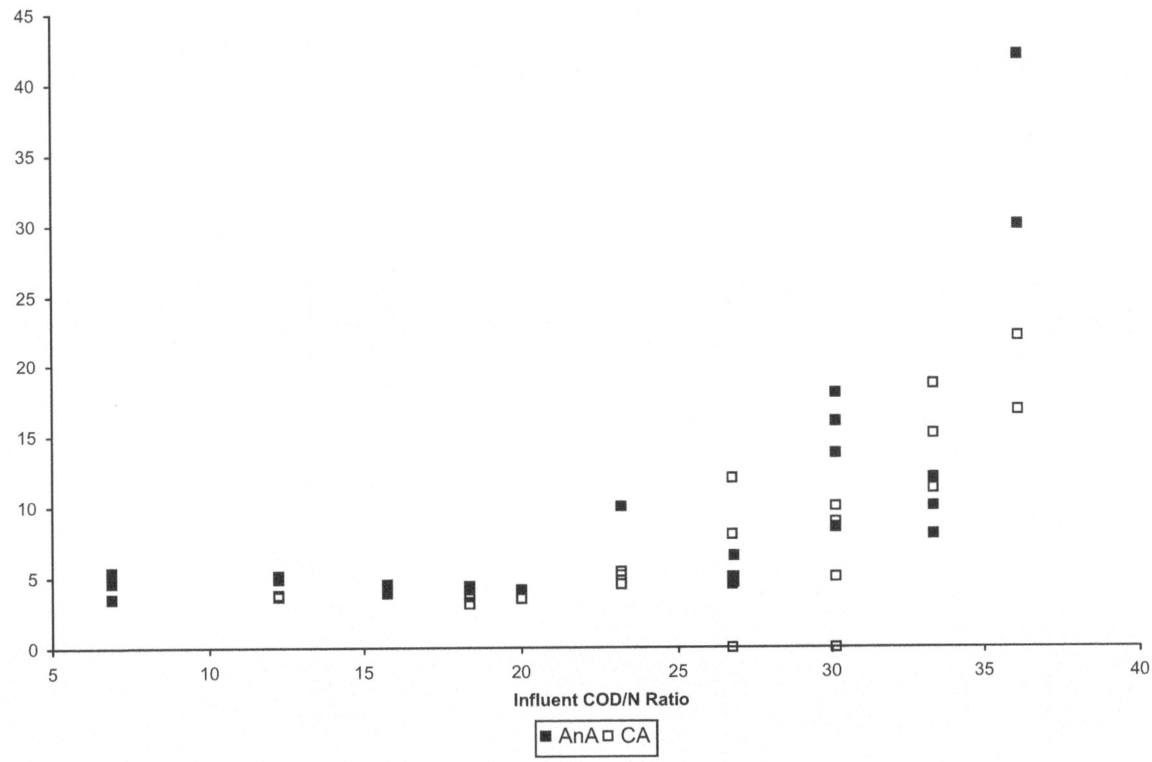

Figure 3-26. The Effect of Influent COD/N Ratio on Effluent Soluble COD Concentration from an AnA SBR and a CA SBR.

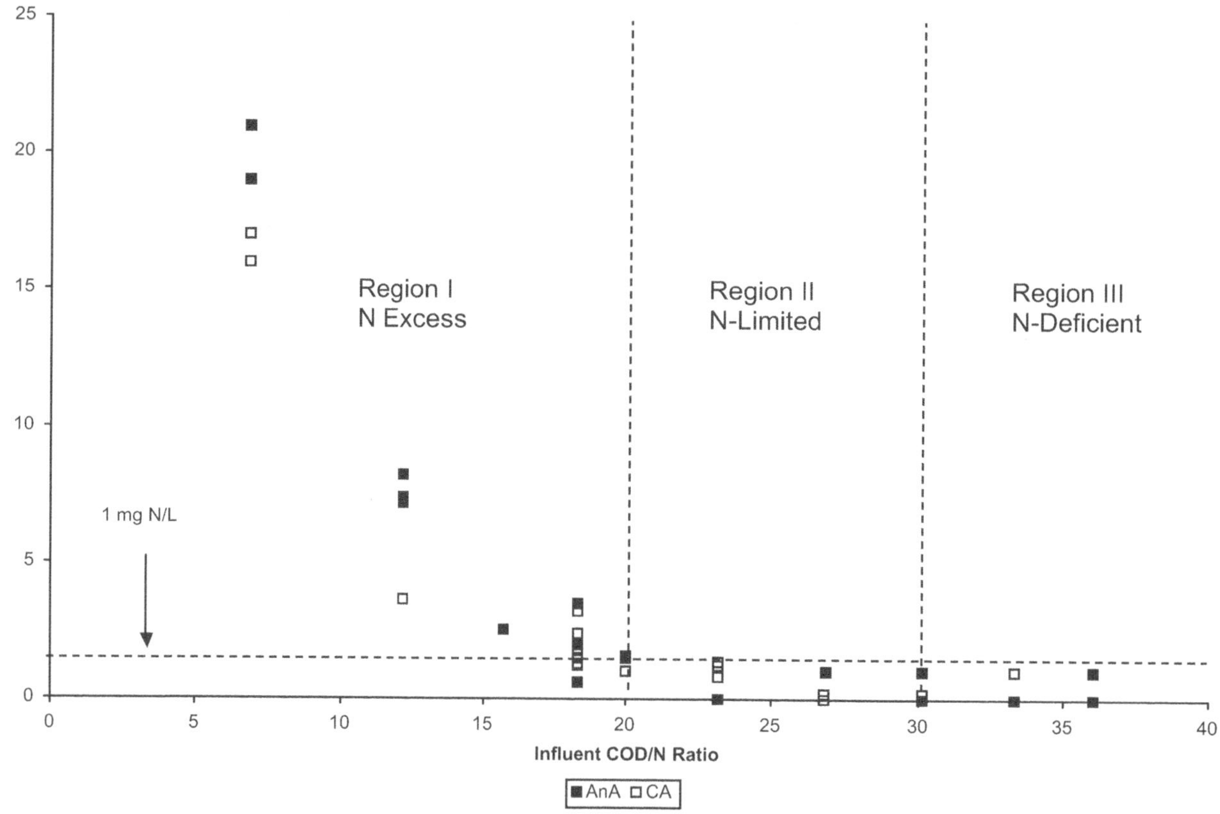

Figure 3-27. The Effect of Influent COD/N Ratio on Effluent Total Soluble Inorganic N Concentration from an AnA SBR and a CA SBR.

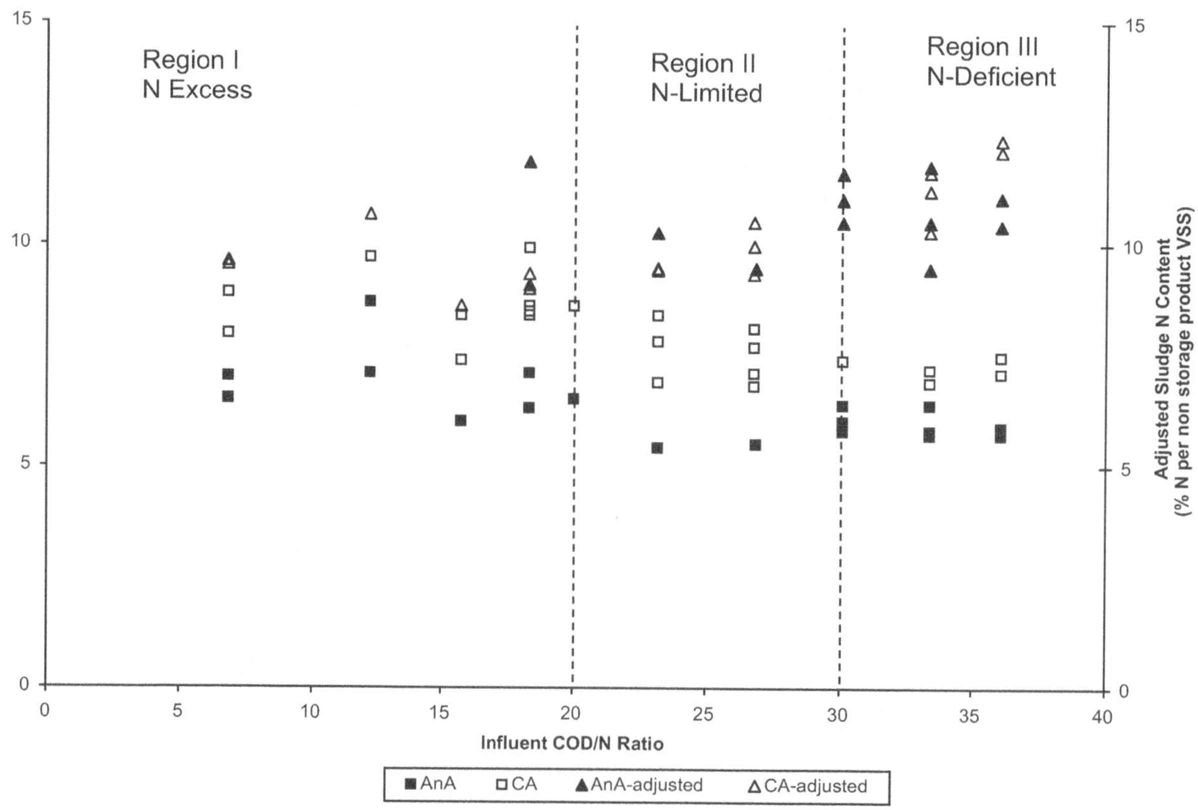

Figure 3-28. The Effect of Influent COD/N Ratio on Sludge N Content in an AnA and a CA SBR.

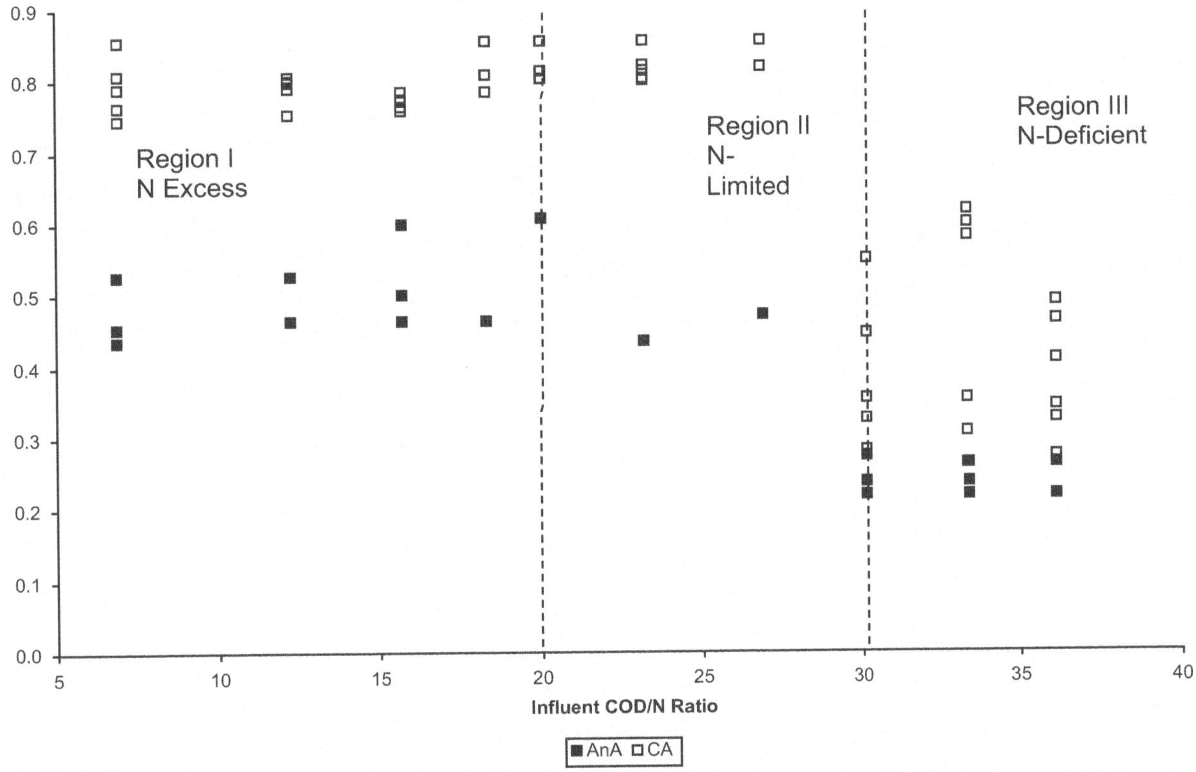

Figure 3-29. The Effect of Influent COD/N Ratio on Acetate Removal Rates During the First 110 Minutes of the Cycle of an AnA and a CA SBR.

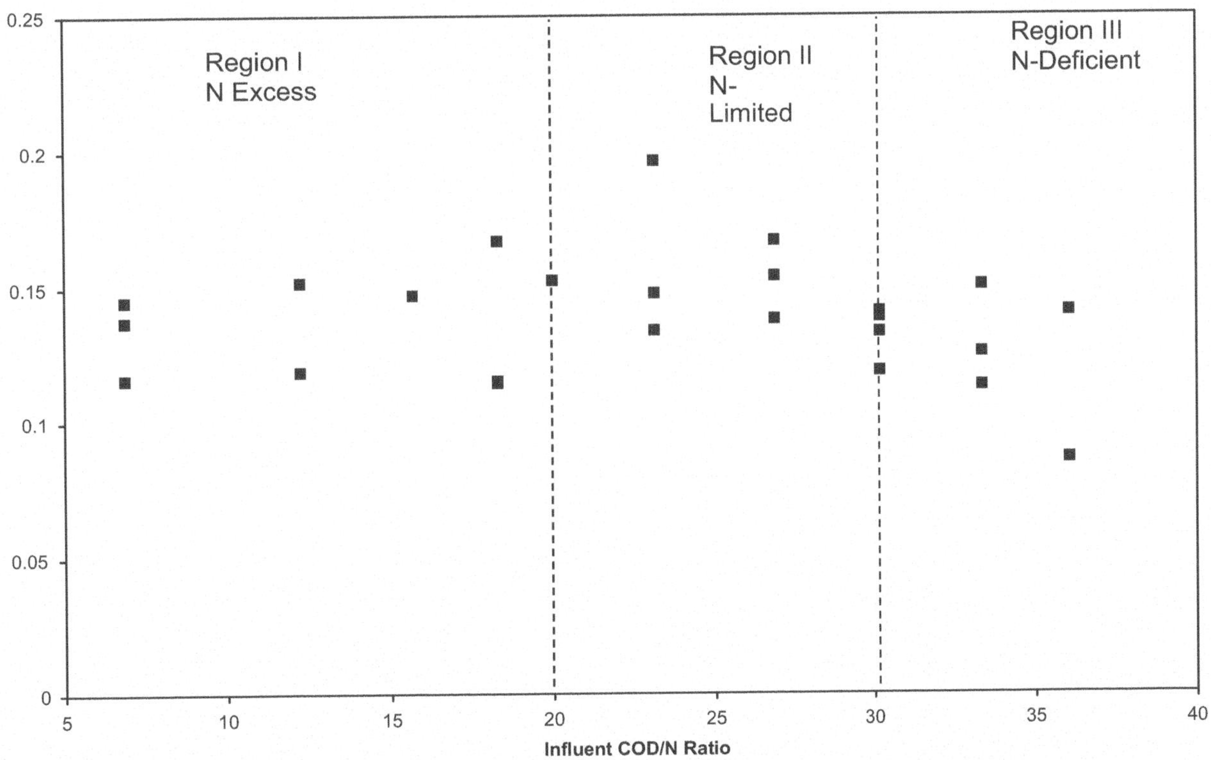

Figure 3-30. The Effect of Influent COD/N Ratio on P Released/Acetate-Carbon Removed Ratio of an AnA SBR.

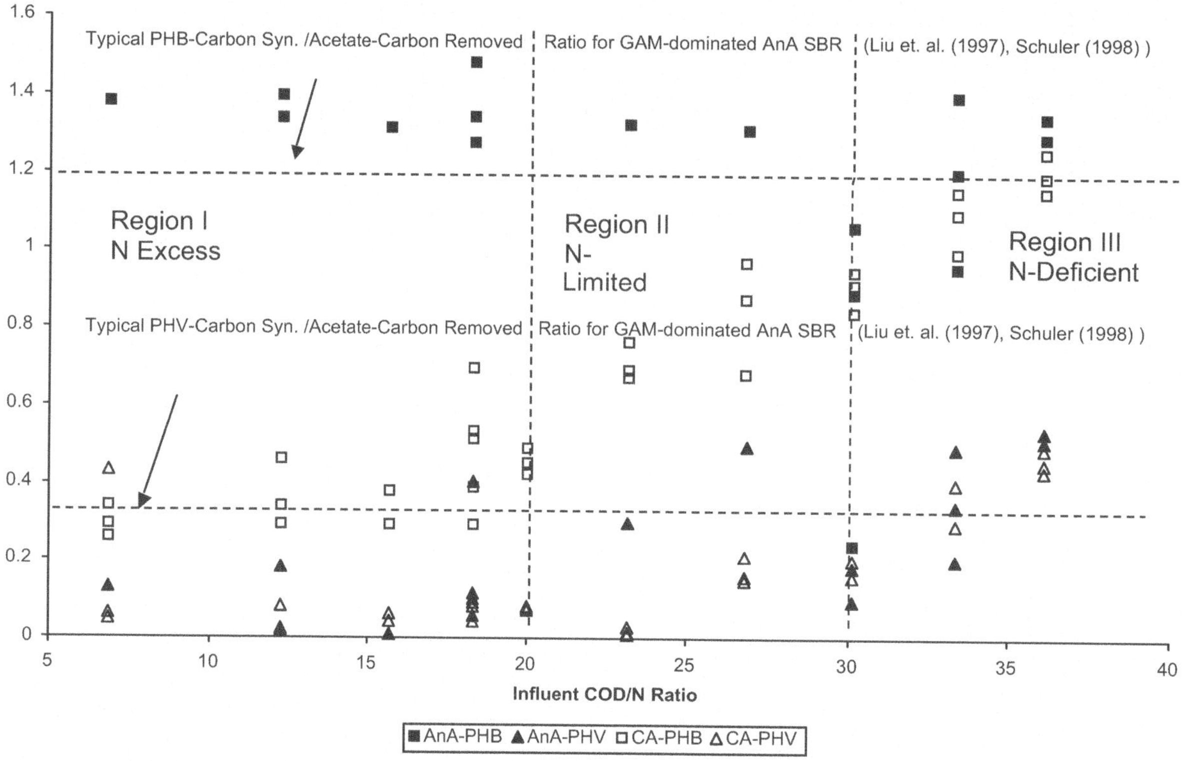

Figure 3-31. The Effect of Influent COD/N Ratio on PHA-Carbon Synthesized/Acetate-Carbon Removed Ratio During the First 110 Minutes of the Cycle of an AnA and a CA SBR.

PHV synthesis increased in both sludges, PHB synthesis in the CA sludge increased, and the fraction of PHB synthesized in the AnA sludge was relatively constant. In Region III, the PHB-C synthesized/acetate-C and PHV-C synthesized/acetate-C ratios for both sludges converged to approximately 1.2 and 0.40, respectively. The AnA PHB-C synthesized/acetate-C and PHV-C synthesized/acetate-C synthesis ratios were generally in good agreement with what would be expected of an AnA system dominated by the GAM. The influent COD/N ratio had no effect on the CH-C degraded/acetate-C stoichiometry for the AnA SBR, but the CH-C degraded/acetate-C removed ratio increased linearly in the CA SBR (Figure 3-32). This finding is consistent with the increased synthesis of PHA in the CA SBR because CH degradation provides reducing equivalents (i.e., NADH) required for PHA synthesis. These results show that N-limiting conditions caused more PHB synthesis in the CA sludge, and more PHV synthesis in both sludges.

3.3.6 End-of-aerobic Storage Product Content, SPY, and Non-storage Product Yield

The AnA PHA content increased from approx 5–25% as the influent COD/N ratio was increased from 7 to 30 (Figure 3-33). PHA was virtually absent from the CA sludge when N was in excess, but it increased significantly when N became limiting. At the N deficiency point, the difference in the PHA content between the sludge types was small (approximately 2%).

The AnA biomass CH content was relatively constant (approximately 20–25% on average) over the range of influent COD/N ratios tested (Figure 3-34). The CA biomass CH content increased from approximately 10% of VSS at an influent COD/N ratio of 7% to approximately

17% of VSS at an influent COD/N ratio of 36. At the N deficiency point (influent COD/N ratio of 30), the difference in the sludge CH content was small (approximately 3% of VSS). The exo-cellular CH content of both reactors was similar and increased as the influent COD/N ratio was increased (Figure 3-35). The intracellular CH content of the AnA decreased, while that of the CA sludge remained constant.

These results show that N limitation increased PHA accumulation in both types of sludge, so that PHA became an important sink for influent carbon (and thus influent COD) for both SBRs under N-limited conditions. The CH content of the AnA sludge was relatively constant for all influent COD/N ratios tested, whereas that of the CA sludge increased as the N loading was decreased. The exocellular CH content for both sludges increased as the influent COD/N ratio increased.

The storage product yield for both types of sludge increased as a result of the changes in PHA and CH contents. Figure 3-36 shows that, as a result of N limitation and N deficiency, the storage product yield of the CA sludge approached that of the AnA sludge such that, when N became deficient, the difference in the amount of non-N containing storage product was small. The non-storage product yield for both sludges was similar, and only declined slightly as N deficiency was approached (Figure 3-37).

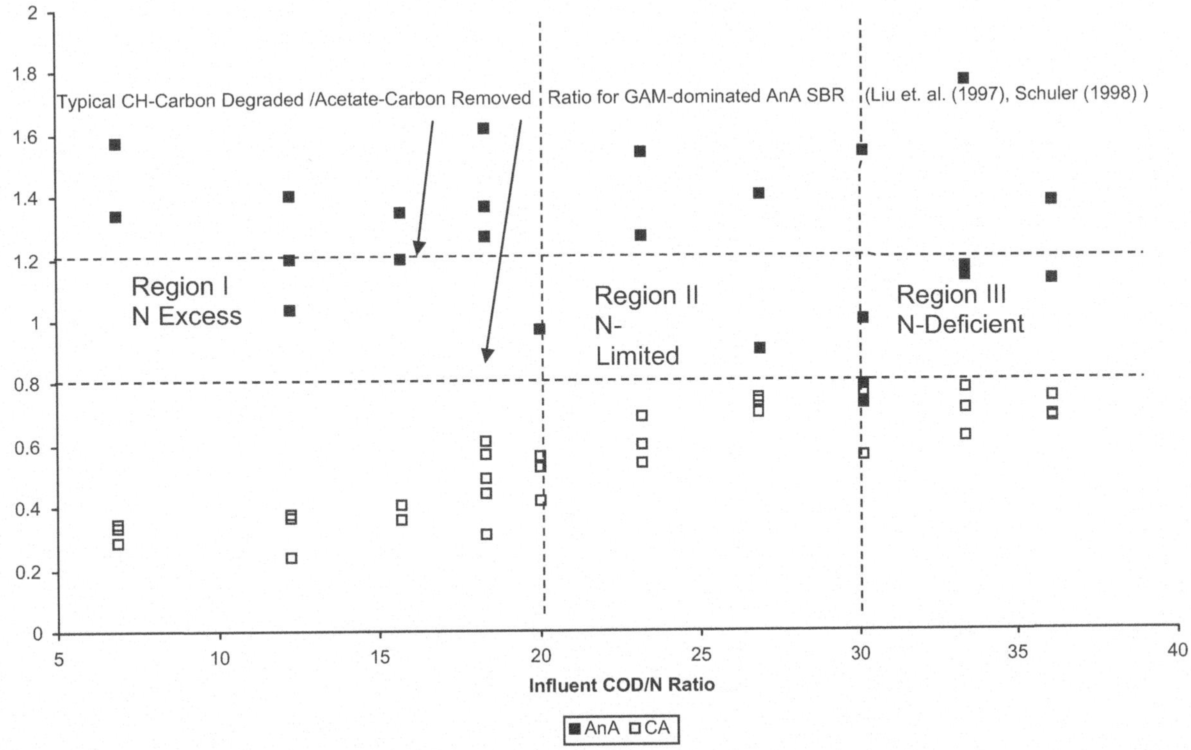

Figure 3-32. The Effect of Influent COD/N Ratio on CH-Carbon Degraded/Acetate-Carbon Removed Ratio During the First 110 Minutes of the Cycle of an AnA and a CA SBR.

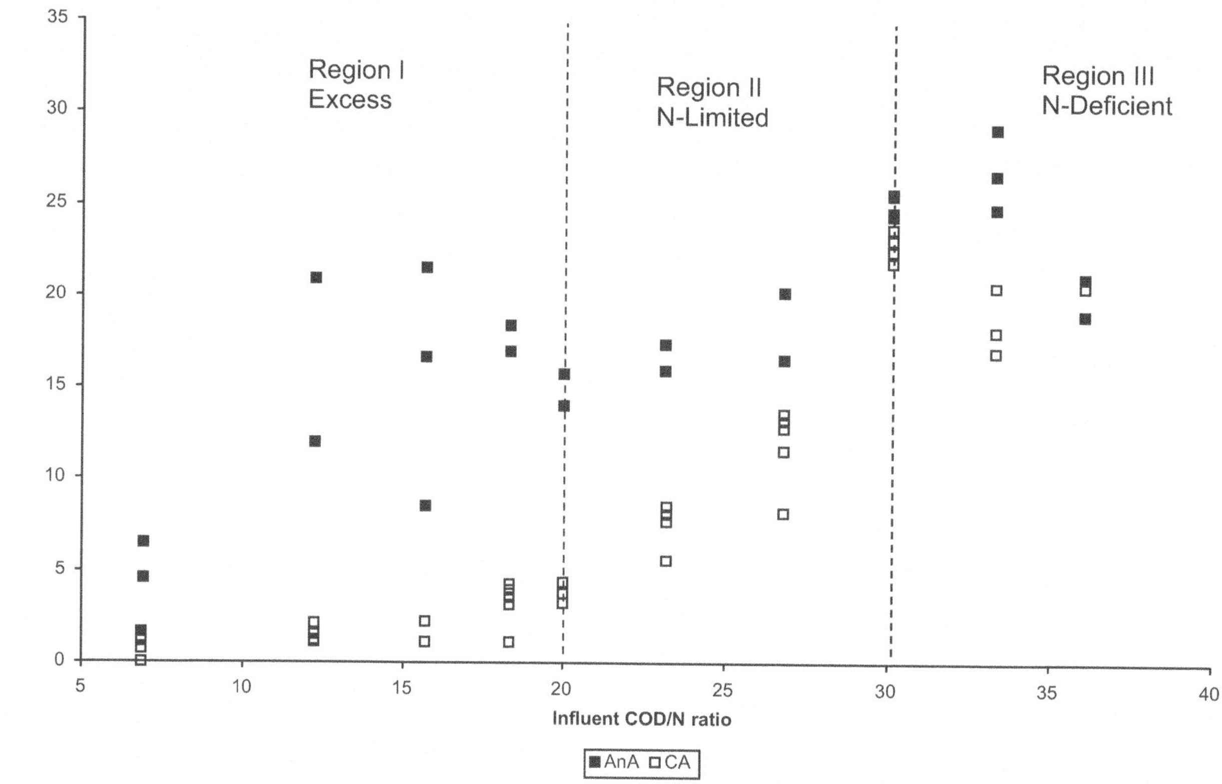

Figure 3-33. The Effect of Influent COD/N Ratio on EOA Biomass PHA Content of an AnA SBR and a CA SBR.

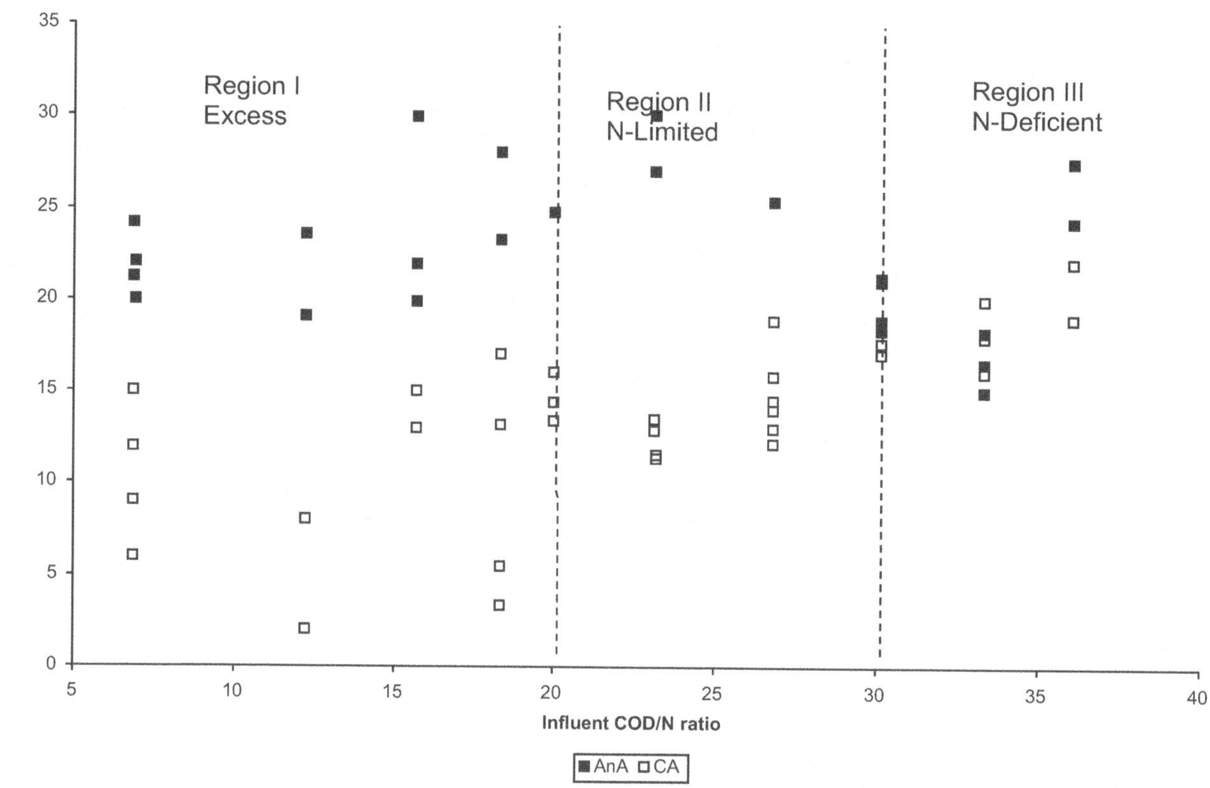

Figure 3-34. The Effect of Influent COD/N Ratio on EOA Cycle Biomass CH Content of an AnA SBR and a CA SBR.

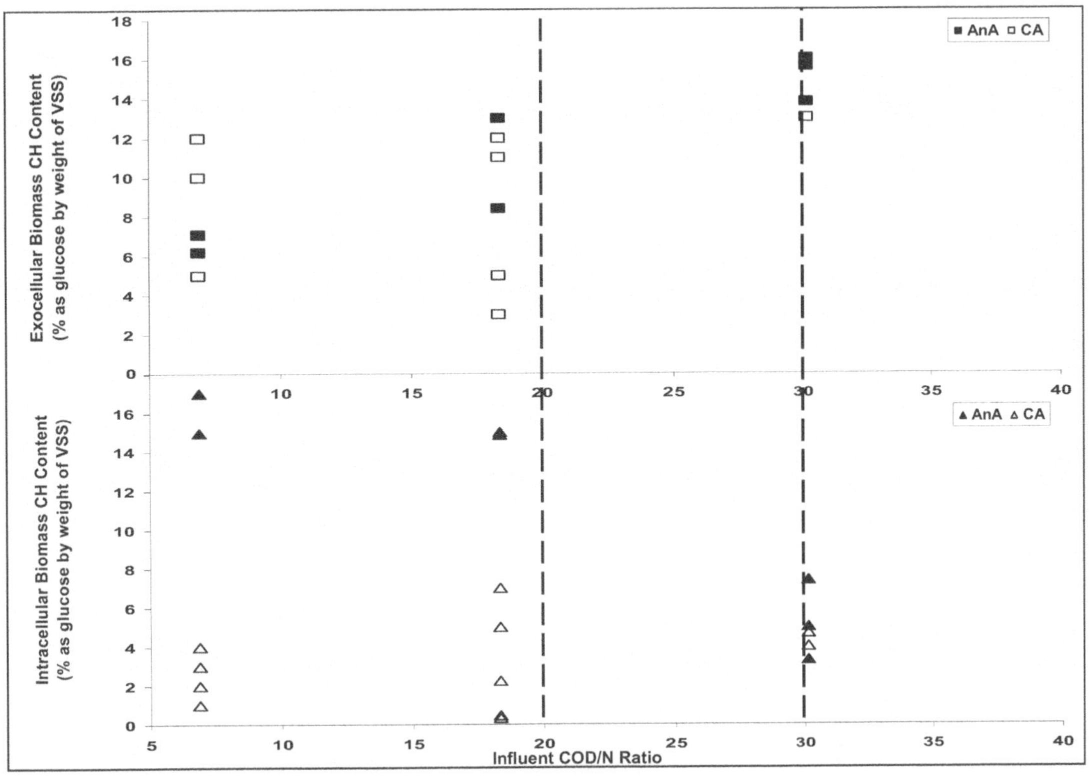

Figure 3-35. The Effect of Influent COD/P Ratio on Exocellular and Intracellular CH Content of an AnA and a CA SBR.

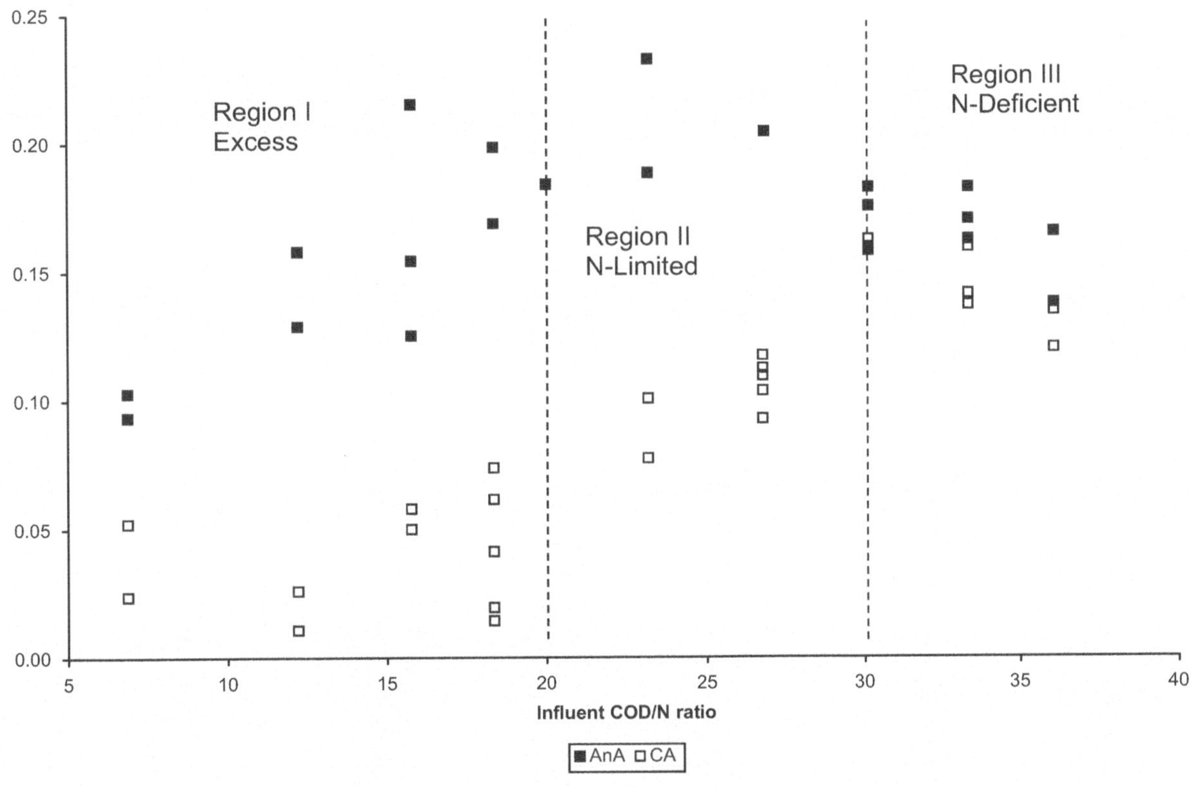

Figure 3-36. The Effect of Influent COD/N Ratio on Biomass Storage Product Yield of an AnA and a CA SBR.

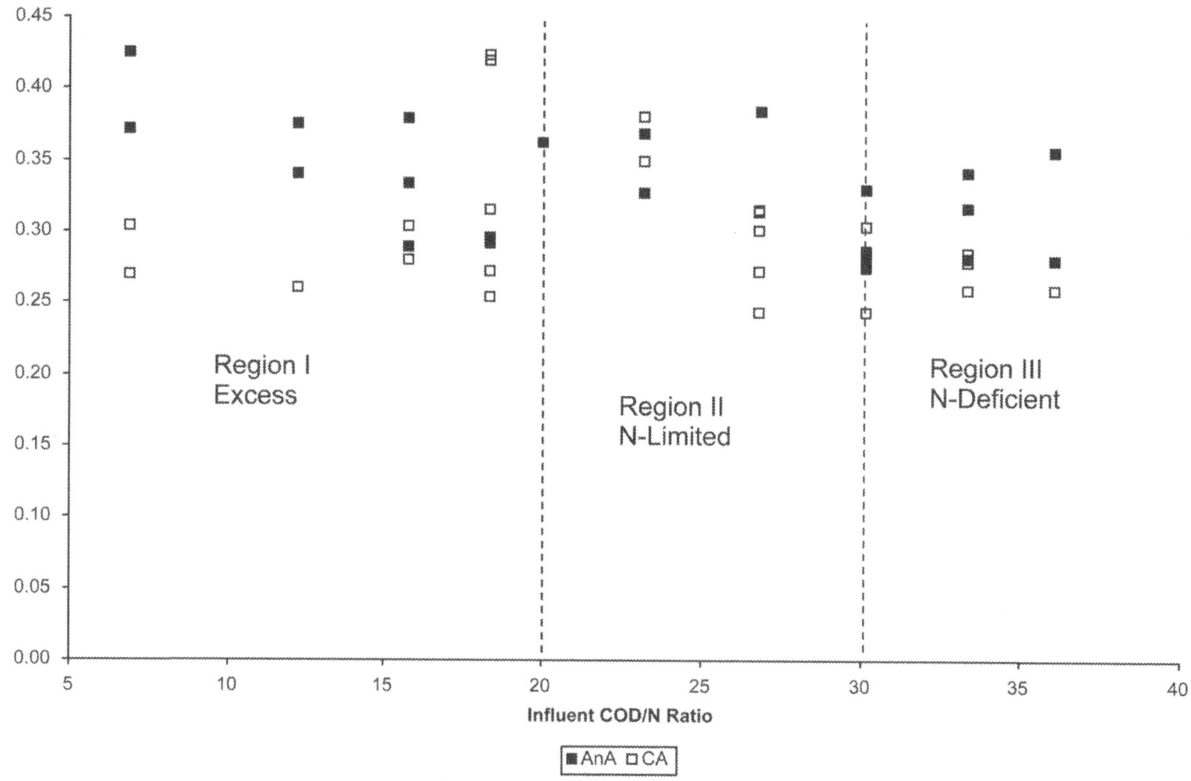

Figure 3-37. The Effect of Influent COD/N Ratio on Biomass Non-Storage Product Yield of an AnA and a CA SBR.

The behavior of the AnA and CA activated sludge with respect to N limitation differed from that for P limitation. There was no discernable difference in the influent COD/N ratio at which N limitation (effluent total inorganic N <1.0 mg N /L) and N deficiency (effluent acetate concentration ≥5 mg acetic acid /L) occurred (Figures 3-25 and 3-27). These results were consistent with the small differences in sludge PHA and CH contents of the AnA and CA SBRs. At the point of N deficiency (influent COD/N ratio = 30), the difference in total non-N containing storage product content of the AnA and CA activated sludges was approximately 5% of VSS, which is consistent with the finding that both sludge types became N-deficient at the same influent COD/N ratio (Table 3-6). From a practical point of view, these data suggest that the use of AnA activated sludge will not provide any advantage over CA activated sludge with respect to steady-state N requirements.

Table 3-6. Conditions at N-Deficiency.

Reactor	PHA (%VSS)	CH (% VSS)	Storage Product COD (%VSS)	Influent COD/N at N deficiency
AnA	25	20	57	30
CA	23	17	52	30
Difference	2	3	5	0

3.3.7 SVI and Effluent TSS

The SVI of the AnA and CA sludge increased when the influent COD/N ratio exceeded 30, but the effluent TSS concentrations were relatively constant for all influent COD/N ratios (Figure 3-38). These data indicate that the presence of the anaerobic zone did not provide any benefit with respect to sludge settling or suspended solids removal.

3.3.8 Carbon Balance

Carbon balances were performed over the first 110 min and the entire cycle for all influent COD/N ratios tested as described in Section 3.2.8. The carbon recoveries for the AnA SBR were higher than those of the CA SBR for both the initial 110 min and the entire cycle (Figure 3-39). The carbon recoveries for the AnA SBR ranged from 94–100%, while those of the CA SBR ranged from 84–93%.

3.3.9 COD Balance

COD recoveries over the first 110 min of the cycle were ≥95% for the AnA and CA reactors (Figure 3-40). COD recoveries over the entire cycle were ≥90% and 88% for the AnA and CA SBR, respectively. The lowest COD recoveries were obtained for both SBRs when acetate as detected in the effluent.

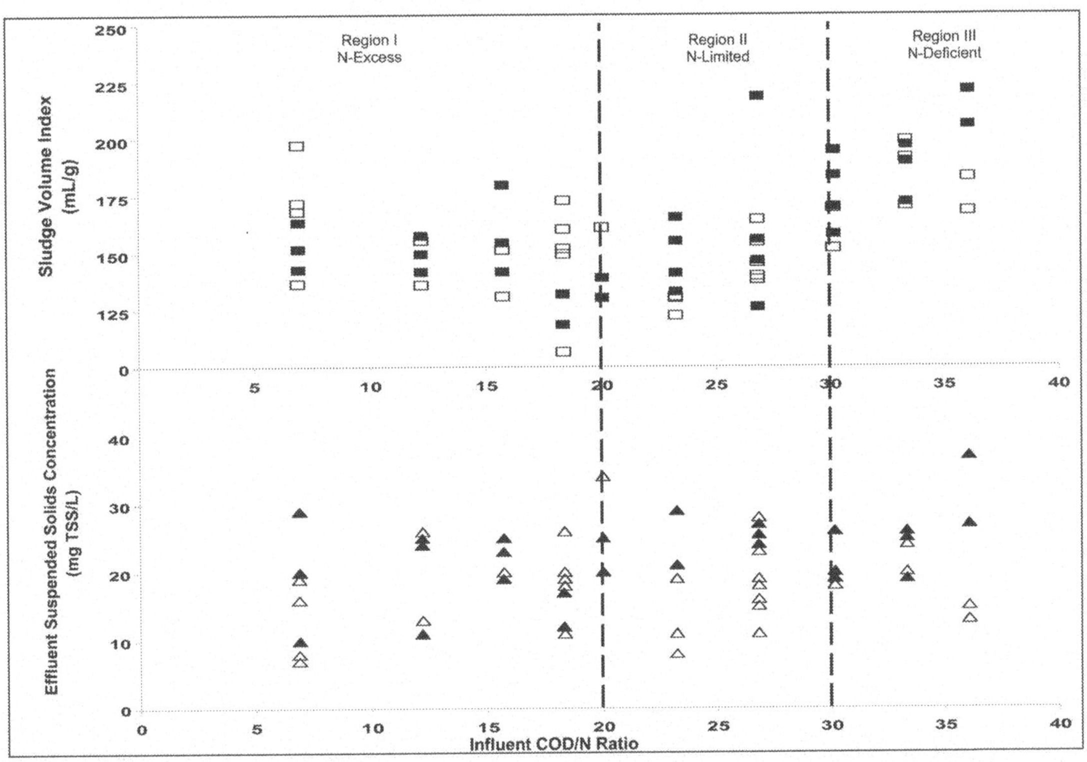

Figure 3-38. The Effect of Influent COD/N Ratio on SVI and Effluent Suspended Solids Concentration for an AnA SBR and a CA SBR.

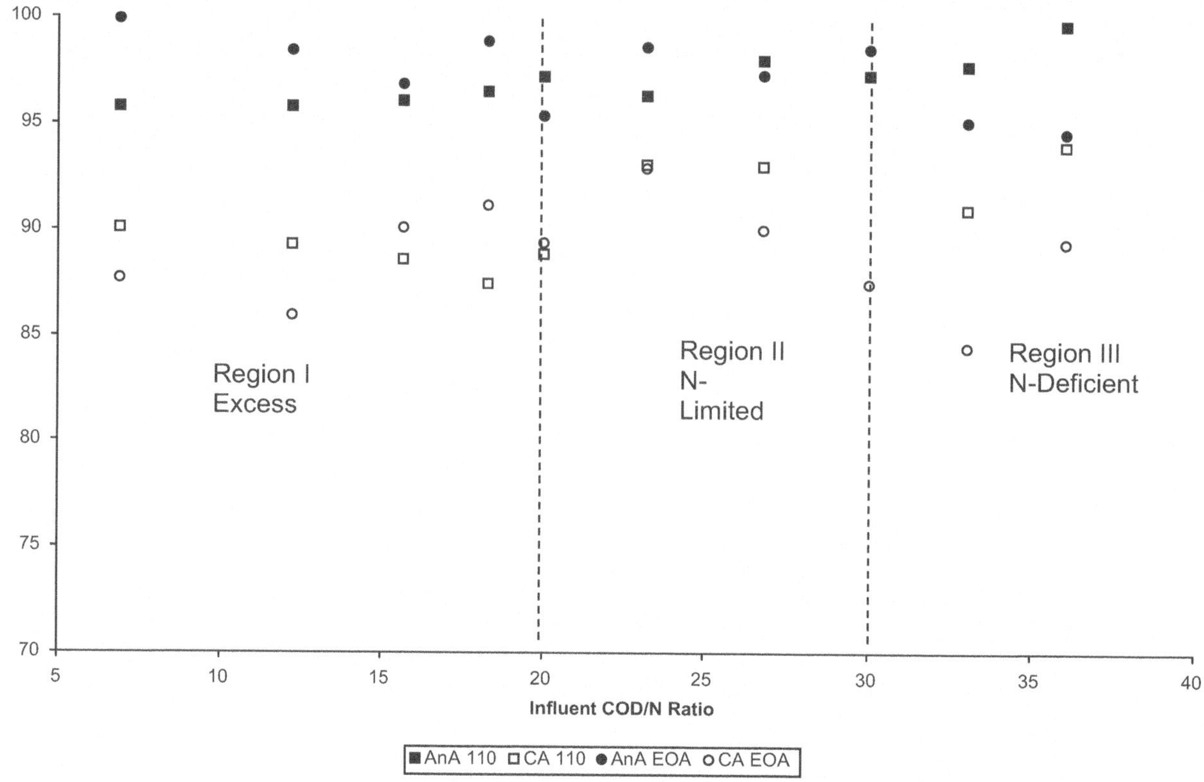

Figure 3-39. The Effect of Influent COD/N Ratio on Carbon Recovery for an AnA SBR and a CA SBR.

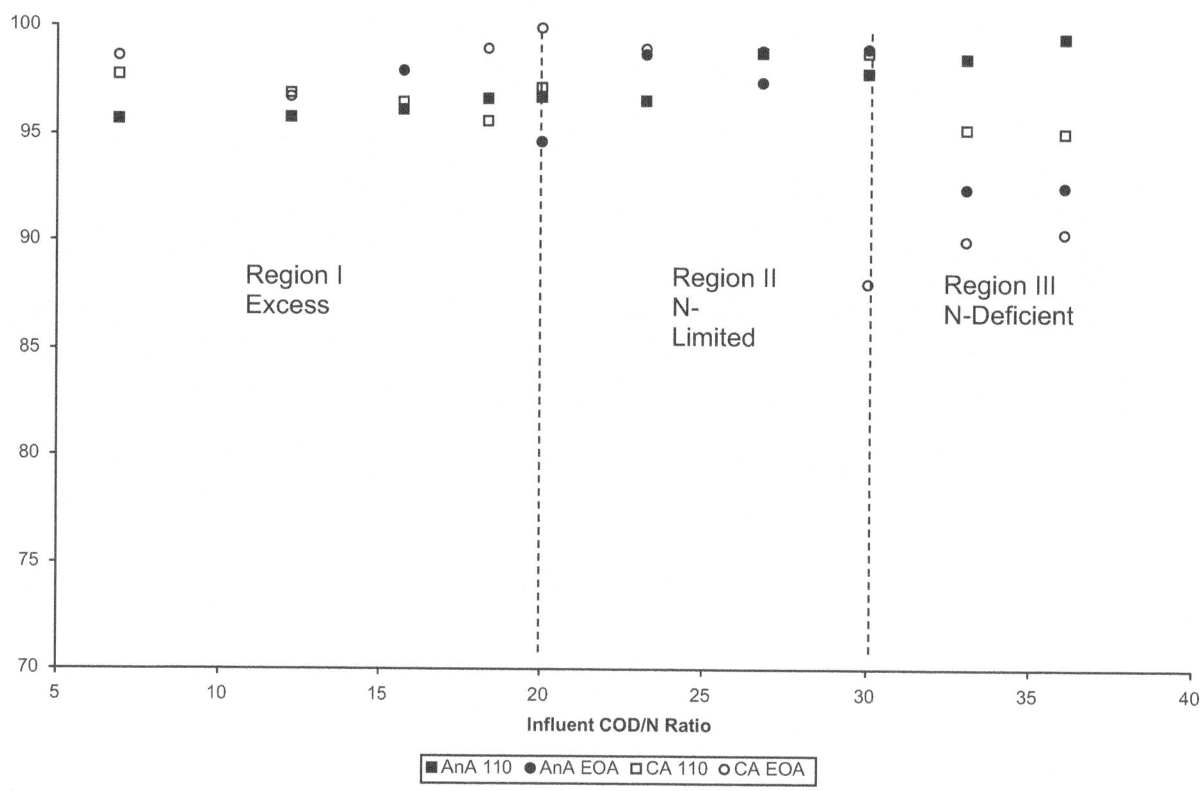

Figure 3-40. The Effect of Influent COD/N Ratio on COD Recovery for an AnA SBR and a CA SBR.

3.3.10 Conclusions

Measurements of soluble constituents and biomass storage products from an AnA and a CA SBR were validated with mass balances on N, C, and COD during the course of N requirement experiments. The N requirement determination showed that the presence of an anaerobic zone did not provide an operating benefit with respect to N requirements for activated sludge. Acetate appeared in the effluent streams of the AnA and CA SBR at the same influent COD/N ratio (30). This finding means that at the point of N deficiency the sludges required the same mass of N to remove influent COD. As the influent COD/N ratio was increased, the storage product content of both sludges increased, and was virtually the same at N deficiency, which is consistent with the finding that the sludges had the same N requirement.

N limitation did not affect the acetate uptake rates of either the AnA or the CA sludge, but N deficiency decreased in the acetate uptake rates of both sludges by 50%. The P release/acetate-C removed ratio of the AnA and CA sludge was unaffected by N limitation but decreased slightly during N deficiency. The PHB-C synthesized/acetate-C removed ratio (CA only) and the PHV-C synthesized/acetate-C removed ratio of both types of sludge increased under N-limiting conditions, which supports previous research showing that PHA accumulation is induced by N limitation (Kustu et. al., 1979). The CH-C degraded/acetate-C removed ratio for the CA sludge increased under N-limiting and N-deficient conditions, but that of the AnA sludge was constant.

Investigators also reached the following conclusions:

- N limitation increased the exocellular CH content of both sludges.
- N deficiency caused an increase in the SVI of both sludges, but there were no differences in the SVIs of the AnA and CA sludges under the same N loading conditions. The effluent TSS concentrations from the AnA and CA SBRs were similar for all influent COD/N ratios tested.
- The sludge N content was consistent with what is expected of bacterial cells and activated sludge, whether N was in excess or limiting.
- The stoichiometries of anaerobic PHA synthesis and CH degradation indicate that the GAM was dominant in the AnA SBR.

3.4 Single-cycle Transient COD Loading Experiments and Model Predictions

Experiments were conducted to determine the effect of single-cycle COD loading increases (or decreases) on AnA and CA SBRs. In the first set of experiments, the magnitude of the COD loading transient was increased (COD upshift) until effluent acetate was detected. In the second set of transient COD loading experiments, the magnitude of the COD loading transient was decreased (COD downshift) until the COD loading transient was zero (no COD loading for the cycle). The purpose of these experiments was to determine whether the presence of the anaerobic zone affected the maximum increase in transient COD loading that would not cause elevated effluent acetate concentrations and the maximum decrease in transient COD loading that would not cause elevated effluent P concentrations.

3.4.1 Effluent Acetate and Soluble COD

Acetate was detected in the effluent streams of both the AnA and CA SBR when the influent COD peak loading factor (e.g., the ratio of the transient COD loading to the steady state average COD loading) was 2.6 (Figure 3-41). The acetate breakthrough point occurred between influent COD peak loading factors of 2.3 and 2.6. While the point of effluent acetate breakthrough was predicted well, the AnA and CA models overpredicted the effluent acetate concentration. As expected, the breakthrough of acetate increased the effluent soluble COD (Figure 3-42). These results show that the presence of an anaerobic zone did not affect the capacity of an SBR activated sludge system to remove acetate during single-cycle transient COD upshift loading.

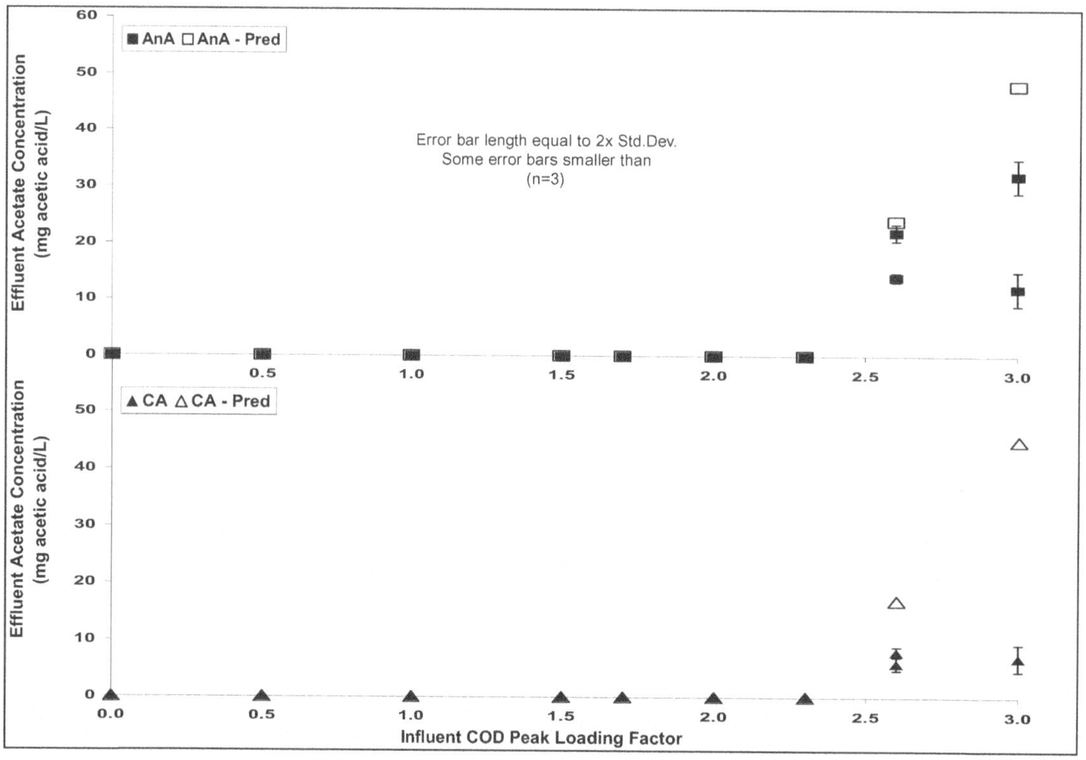

Figure 3-41. The Effect of Influent COD Peak Loading Factor on Effluent Acetate Concentration for an AnA SBR and a CA SBR.

3.4.2 Effluent Soluble Phosphorus

Figure 3-43 shows that the effluent soluble P from the CA was higher than the AnA SBR when the influent COD peak loading factor was ≤1, but when the influent COD peak loading factor was ≥1.5, the effluent P concentration from both reactors was ≤0.5 mg/L (with the exception of one data point from the AnA SBR). The measured effluent AnA and CA P concentrations correspond with the predicted values. This result shows that the AnA reactor produced lower effluent P levels than the CA SBR during low COD loading transients and that both reactors produced low effluent soluble P concentrations during high COD loading transients.

3.4.3 Stoichiometry of P Release, PHA Synthesis, and CH Degradation

The stoichiometry of P release (AnA sludge only), PHA synthesis, and CH degradation was determined during the transient cycle for the AnA and CA sludges. As the influent COD peak loading factor increased, the P release/acetate-C removed ratio increased for the AnA SBR (Figure 3-44), indicating that more P was released per mole of acetate-C removed during transient (upshift) COD loading conditions than at average COD loading conditions.

The PHA-C synthesized/acetate-C removed ratio of the AnA SBR was higher than for the CA SBR (Figure 3-45). The AnA PHA-C synthesized/acetate-C removed ratio was constant and approximately 1.1 mole PHA-C synthesized/mole acetate-C (close to the theoretical ratio of 1.33), while in the CA sludge this value ranged from 0.2 to 1.0 mole PHA-C synthesized/mole acetate-C and increased linearly as the influent COD peak loading factor increased.

As the influent COD peak loading factor was increased, the ratio of CH-C degraded/acetate-C removed in the AnA sludge decreased from approximately 1.1 to 0.8, while in the CA sludge the ratio showed an increasing trend for influent COD peak loading factors <1.0 and a gradually decreasing trend for influent COD peak loading factors >1.0 (Figure 3-46).

The correlations of P release/acetate-C removed, CH-C degraded/acetate-C removed, PHA-C synthesized/acetate-C removed shown in Figures 3-44 through 3-46 were incorporated into the CA and AnA models for modeling of transient COD loading conditions (see Appendix B).

3.4.4 End-of-aerobic VSS, Sludge Storage Product Content

The end-of-aerobic biomass content of the AnA and CA sludge increased linearly as the influent COD peak loading factor increased (Figure 3-47). The PHA content increased gradually when the influent COD peak loading factor was <1.0 then more significantly with influent COD peak loading factors of between 1.0 and 2.0 (Figure 3-48).

The PHA contents of both sludges were relatively constant (neglecting outliers) when the influent COD peak loading factor was >2.0. The AnA CH content increased with influent COD peak loading factor up to 2.0, then decreased between values of 2.0 and 3.0 (Figure 3-49). The CA CH content increased until the influent COD peak loading factor was 1.7, then decreased as the influent COD peak loading factor approached 3.0. The trends shown by the AnA and CA model predictions for VSS, PHA, and CH agree with those of the measured values. The experimental results show that more carbon is retained as PHA and less as CH for influent COD peak loading factors >2.0.

3.4.5 Acetate Uptake Rates

Figure 3-50 shows that the acetate uptake rates of the CA SBR sludge were approximately 25–50% greater than those of the AnA SBR sludge when the influent COD peak loading factor was ≥1.0. As the influent COD peak loading factor increased, the acetate uptake rates appeared to approach a maximum of approximately 1.5 mg/L/min for the CA SBR sludge and 1.1 mg/L/min for the AnA SBR sludge. These maximum acetate uptake rates were reached at influent COD peak loading factors of 2.0–2.6, which is near the point at which acetate appeared in the effluents (Figure 3-41). The measured values agree well with AnA and CA model predictions.

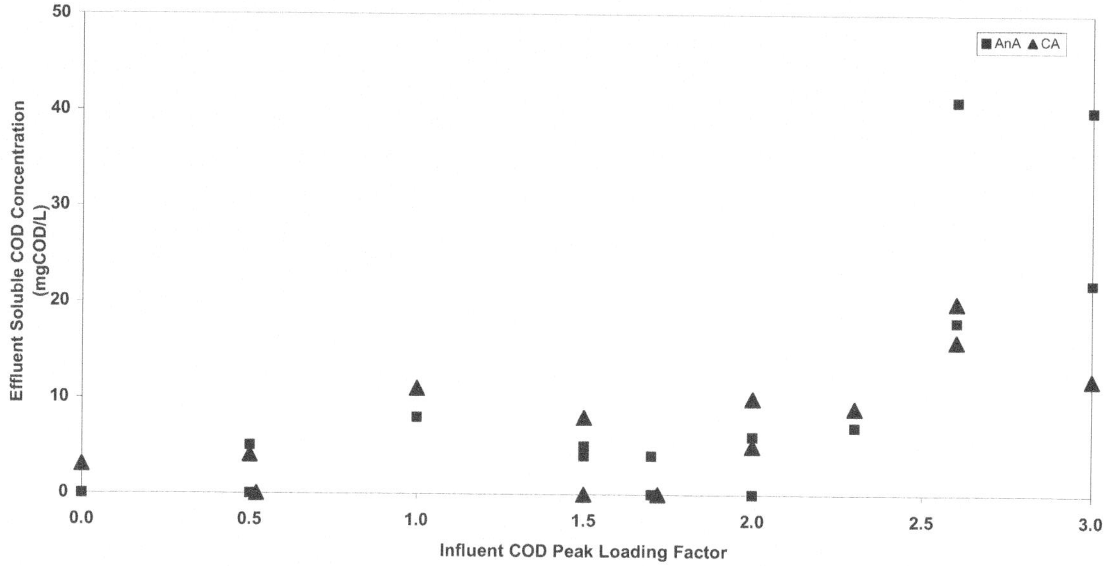

Figure 3-42. The Effect of Influent COD Peak Loading Factor on Effluent Soluble COD Concentration of an AnA SBR and a CA SBR.

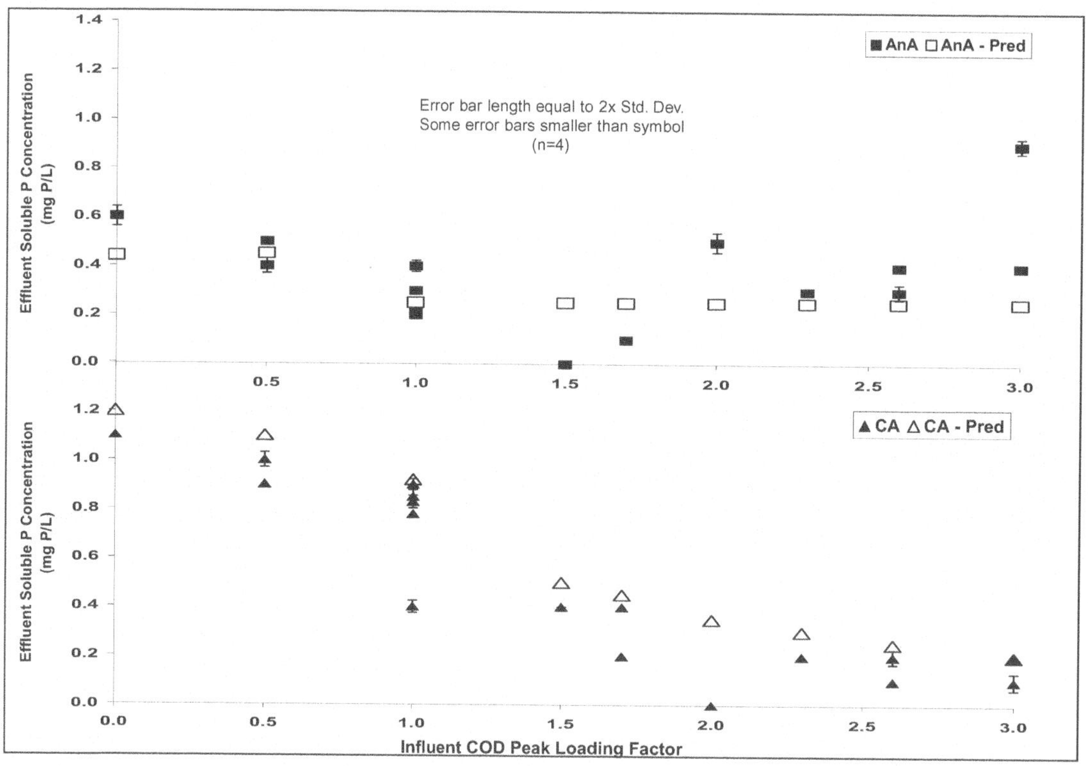

Figure 3-43. The Effect of Influent COD Peak Loading Factor on Effluent Soluble P Concentration for an AnA SBR and a CA SBR.

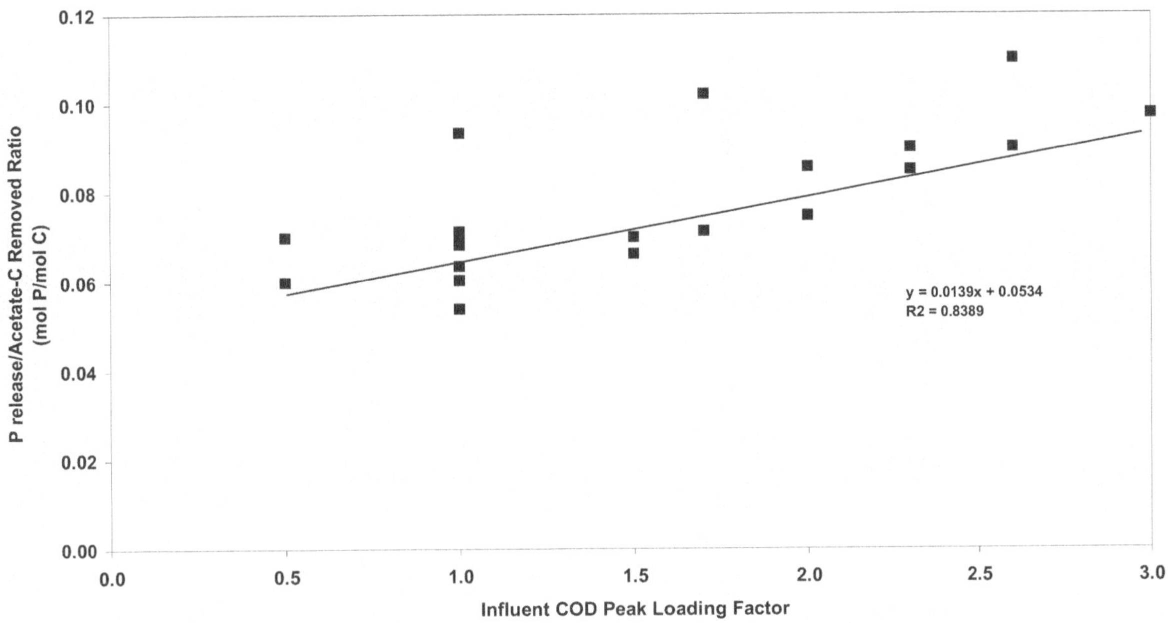

Figure 3-44. The Effect of Influent COD Peak Loading Factor on P release/Acetate-C Removed Ratio of an AnA SBR.

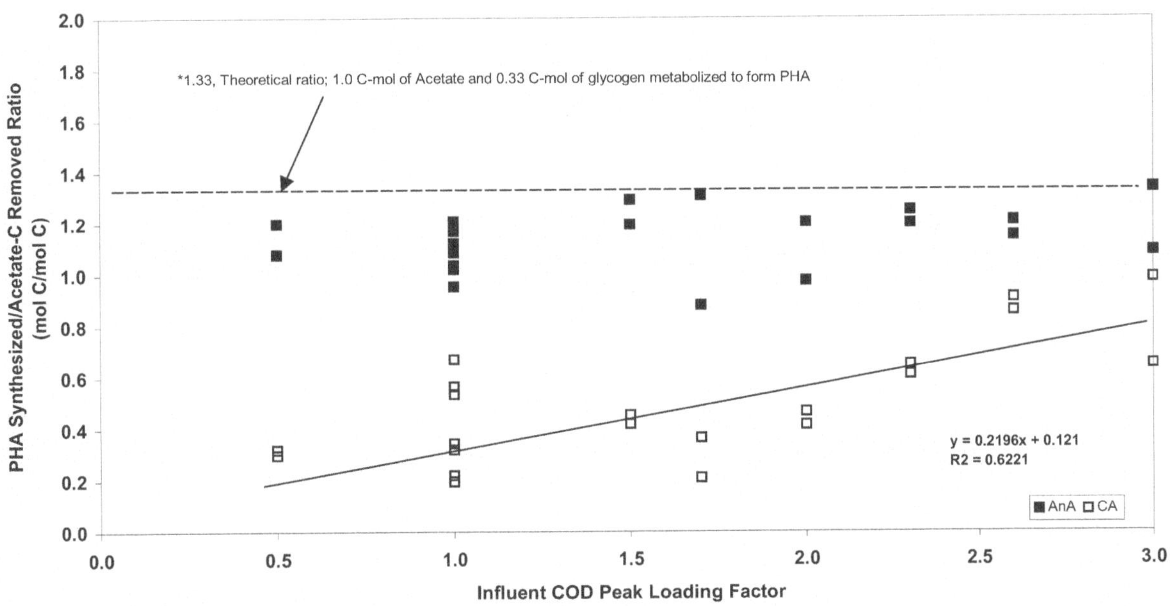

Figure 3-45. The Effect of Influent COD Peak Loading Factor on PHA Synthesized/Acetate-C Removed Ratio for an AnA SBR and a CA SBR.

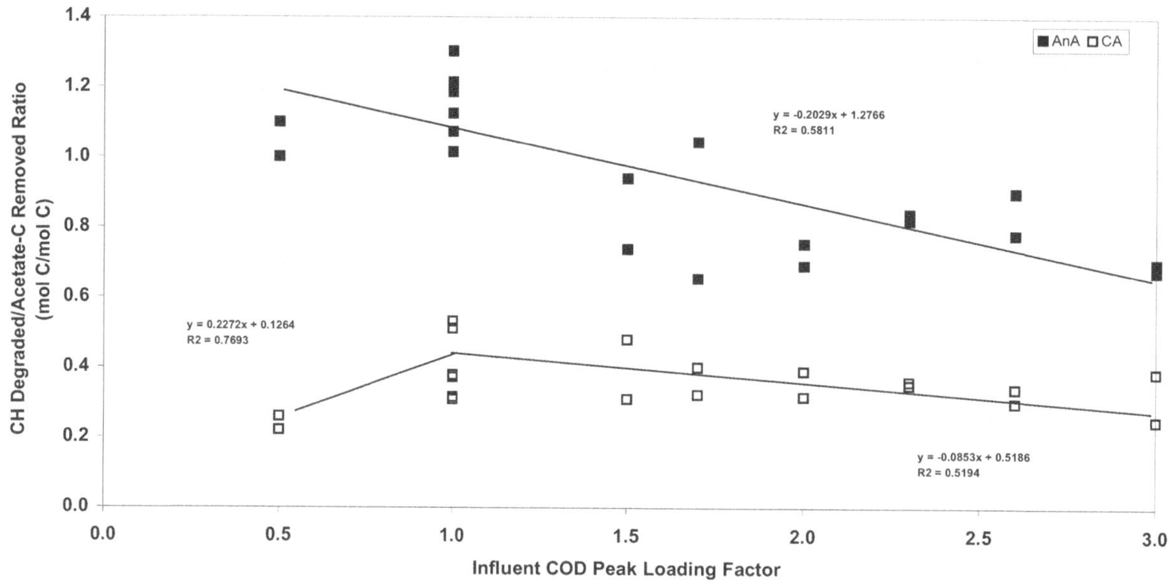

Figure 3-46. The Effect of Influent COD Peak Loading Factor on CH-C Degraded/Acetate-C Removed Ratio for an AnA SBR and a CA SBR.

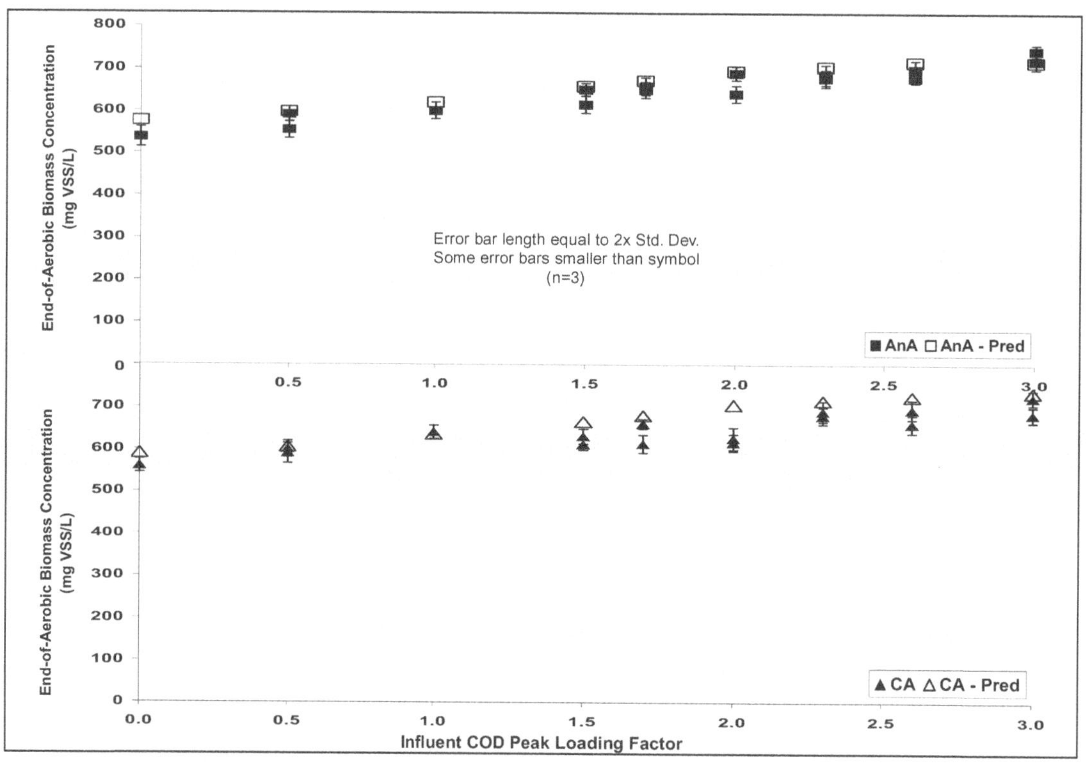

Figure 3-47. The Effect of Influent COD Peak Loading Factor on End-of-Aerobic Biomass Concentration for an AnA SBR and a CA SBR.

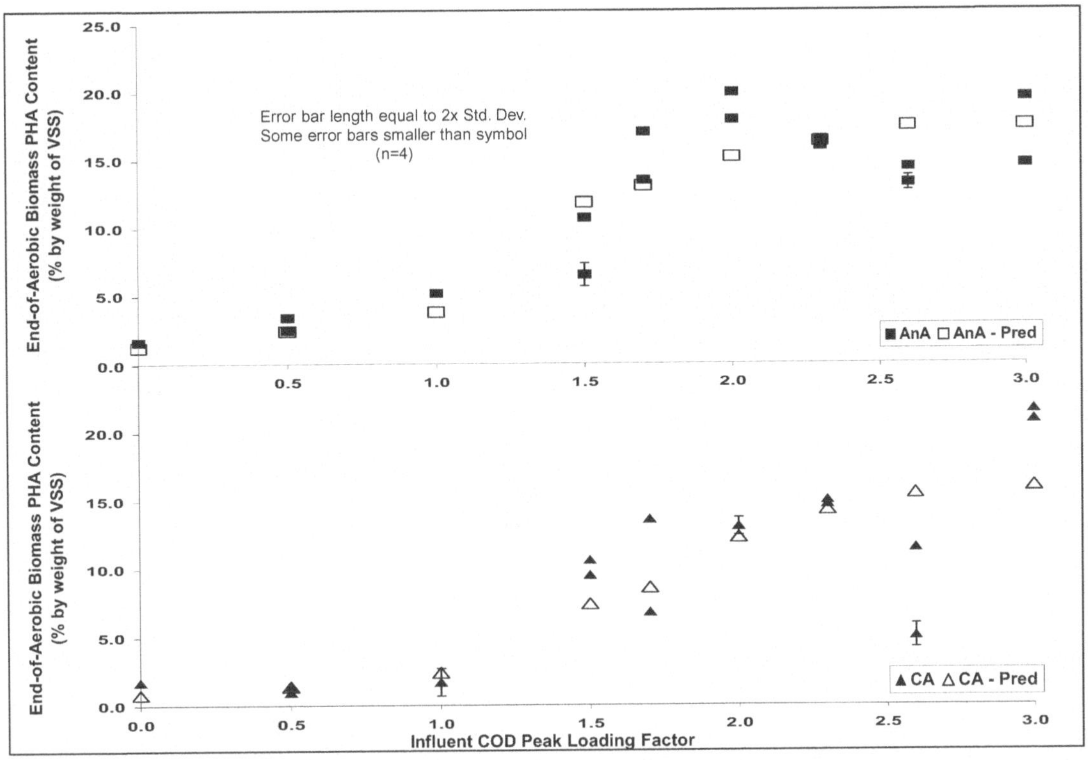

Figure 3-48. The Effect of Influent COD Peak Loading Factor on End-of-Aerobic Biomass PHA Content for an AnA SBR and a CA SBR.

3.4.6 Conclusions

Experiments on laboratory scale SBRs determined whether the presence of an initial anaerobic zone affected effluent quality and sludge composition during single-cycle transient COD loading. The AnA and CA SBRs removed acetate completely during single-cycle COD upshift experiments when the influent COD peak loading factor was ≤ 2.3. When the influent COD peak loading factor was ≥ 2.6, acetate was detected in the effluents of both SBRs. This result showed that the presence of an anaerobic zone did not affect the capacity of an SBR activated sludge system to remove acetate during single-cycle transient COD upshift loading. COD downshift experiments showed that the effluent P concentration from a CA SBR exceeded 1.0 mgP/L when the influent COD peak loading factor was ≤ 0.5. The AnA effluent soluble P concentration was ≤ 0.5 mgP/L for all influent COD peak loading factors tested. This result showed that during single-cycle COD downshift periods, the AnA SBR maintained low soluble P concentrations while the CA SBR produced effluent soluble P levels of ≥1.0 mgP/L.

The VSS contents of both SBRs increased as the influent COD peak loading factor increased. The VSS PHA content in both SBRs increased as the influent COD peak loading factor increased. The VSS CH content of the SBRs decreased when the influent COD peak loading factor was >2.0 for the AnA sludge and > =1.7 for the CA sludge. This means that high COD loading events (e.g. influent COD peak loading factor > 2.0) cause more carbon to be retained as PHA and less as CH.

Acetate uptake rates increased as the influent COD peak loading factor increased. The acetate uptake rate reached a maximum when the influent COD peak loading factor was between 2.0 and 2.6. This corresponds to the point at which acetate was detected in the effluents of both SBRs.

Process stoichiometry was affected by single-cycle transient COD loading. When the influent COD peak loading factor increased, the P release/acetate-C removed ratio increased for the AnA sludge, and the PHA-C synthesized/acetate-C removed ratio increased for the CA sludge, but was constant for the AnA sludge. The CH-C degraded/acetate-C removed ratio for the AnA SBR decreased as the influent COD peak loading factor increased, while in the CA sludge the ratio showed an increasing trend for influent COD peak loading factors <1.0 and a gradually decreasing trend for influent COD peak loading factors >1.0.

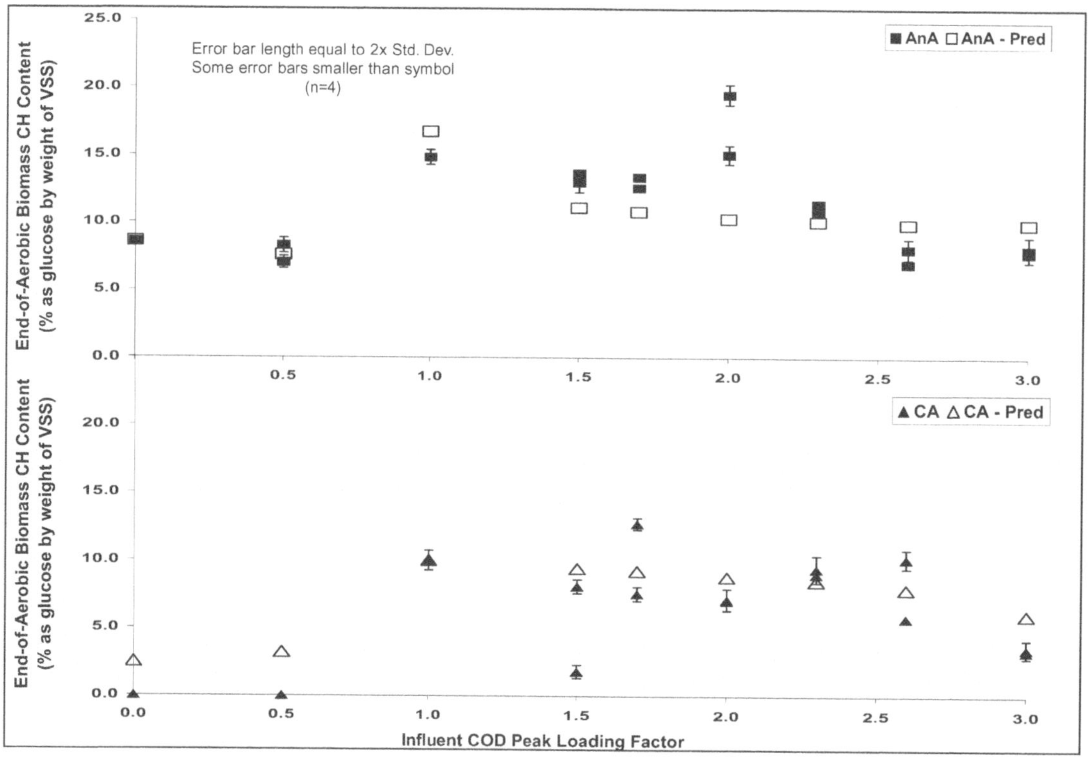

Figure 3-49. The Effect of Influent COD Peak Loading Factor on End-of-Aerobic Biomass CH Content for an AnA SBR and a CA SBR.

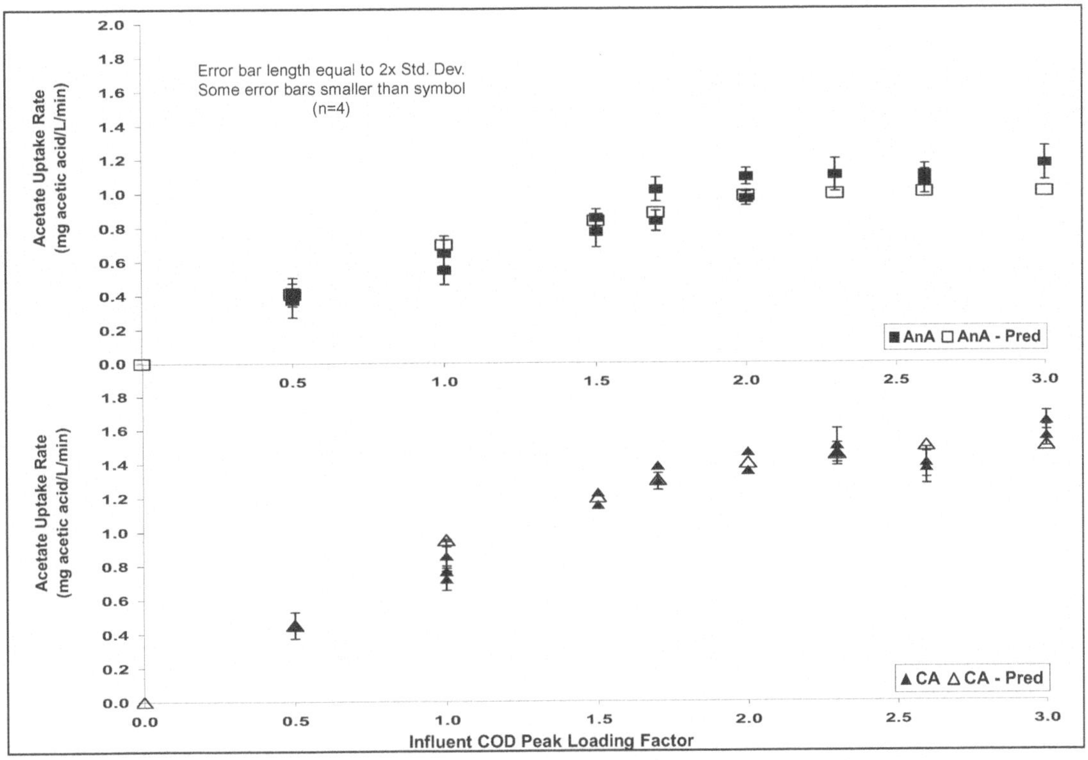

Figure 3-50. The Effect of Influent COD Peak Loading Factor on Acetate Uptake Rate for an AnA SBR and a CA SBR.

3.5 Multiple-cycle Transient COD Loading Experiments and Model Predictions

3.5.1 Influent Loading Patterns

The single-cycle experiments showed that influent COD peak loading factors of ≥2.0 increased the sludge PHA and/or decreased sludge CH levels for both SBRs (Figures 3-48, 3-49), and influent COD peak loading factors of ≤0.5 caused increases in effluent P concentration for the CA SBR but not the AnA SBR (Figure 3-43). To explore the effects of multiple-cycle transients, the daily mass COD loading to the AnA and CA SBRs was added in an unbalanced manner meant to simulate daily COD loading fluctuations (Figure 3-1, Loading Pattern 1). In these experiments, the same total daily mass of COD was added as in the steady-state experiments, the high and low influent COD peak loading factors were 1.8 and 0.2 respectively, and the average overall influent COD/P ratio was 90.

Multiple cycle transient COD loading experiments were also conducted using Loading Patterns 2 and 3 (Figure 3-1). The maximum and minimum influent COD peak loading factors used for the Loading Pattern 2 and 3 experiments were 2.4 and 0.1 respectively. Loading Pattern 2 simulated the effect of a low weekend COD loading period, while Loading Pattern 3 simulated the effect of longer low COD loading periods, such as those that might be caused by holidays or long weekends.

3.5.2 Multiple Cycle Transient COD Loading Results: Loading Pattern 1

3.5.2.1 Effluent P Concentration

Loading Pattern 1 experiments show the AnA SBR maintained lower soluble P concentrations than the CA SBR (Figure 3-51). During the high COD loading events (i.e., influent COD peak loading factor of 1.8), the average soluble P concentrations of the AnA and CA SBRs were both low—0.3 and 0.4 mgP/L, respectively. However, when the influent COD loading was low (i.e., influent COD peak loading factor of 0.2), the average effluent soluble P concentrations from the AnA remained low (0.4 mgP/L), but that of the CA SBR increased to 1.7 mgP/L. Under average COD loading conditions (i.e. influent COD peak loading factor of 1.0), the average effluent soluble P concentration of the AnA SBR was again low (0.4 mgP/L), while that of the CA SBR was 1.0 mgP/L. Table 3-7 shows the daily average effluent soluble P concentrations for both SBRs. Note that there is no trend either up or down in these values suggesting that the SBRs were operating in a stable mode under this COD loading pattern. Figure 3-51 also shows that the experimental data from both the CA and AnA SBR was predicted well by the model. These results show that the AnA process was able to maintain a soluble effluent P concentration of well below 1 mgP/L for Loading Pattern 1 but that the CA SBR was unable to do this.

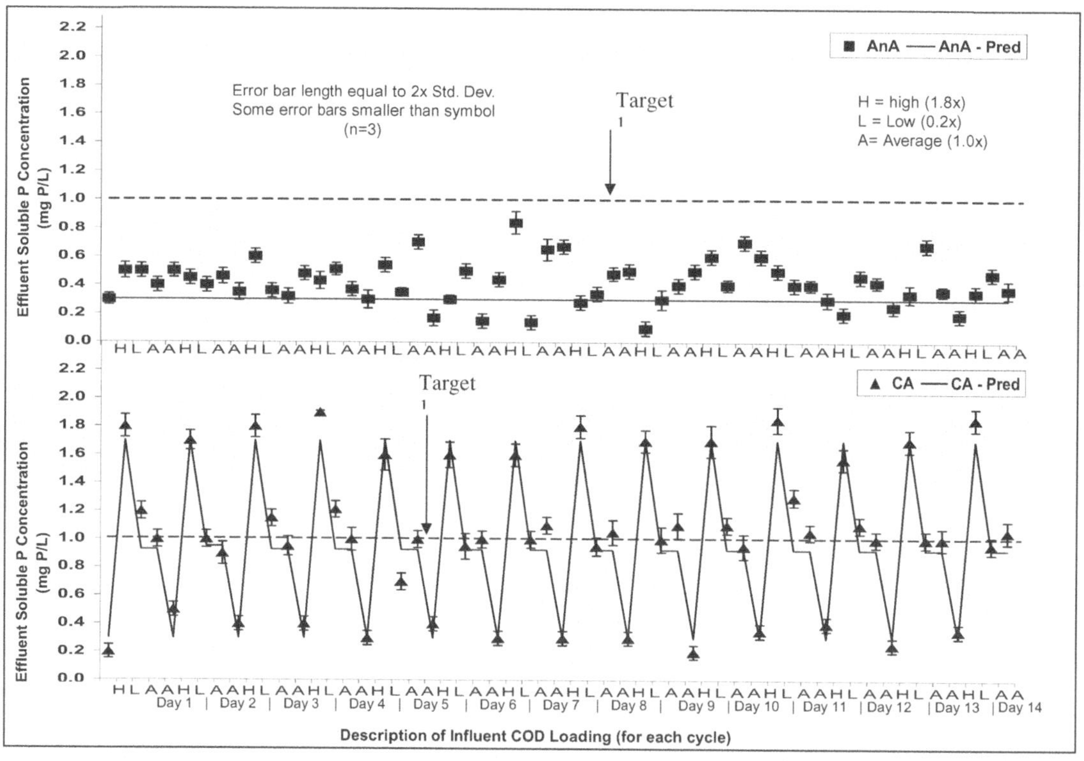

Figure 3-51. The Effect of Variable Influent COD Loading on Effluent Soluble P Concentration from an AnA SBR and a CA SBR (Loading Pattern 1).

Table 3-7. Daily Average Effluent Soluble P Concentration Loading Pattern 1 Experiment.

Day (Pattern)	AnA, mg P/L	CA, mg P/L
1 (H-L-A-A)	0.4	1.0
2 (H-L-A-A)	0.5	1.0
3 (H-L-A-A)	0.4	1.1
4 (H-L-A-A)	0.4	1.1
5 (H-L-A-A)	0.5	0.9
6 (H-L-A-A)	0.3	1.0
7 (H-L-A-A)	0.5	1.0
8 (H-L-A-A)	0.4	1.0
9 (H-L-A-A)	0.3	1.0
10 (H-L-A-A)	0.6	1.0
11 (H-L-A-A)	0.5	1.1
12 (H-L-A-A)	0.3	1.0
13 (H-L-A-A)	0.4	1.0
14 (H-L-A-A)	0.3	1.0
Overall Averages	0.4	1.0
Loading Pattern Symbols: H = 1.8x, A = 1.0x, L = 0.2x		

3.5.2.2 Effluent Acetate and Soluble COD Concentration

The effluent acetate concentration of both reactors throughout these multiple cycle transient COD loading experiments was <2 mg/L (the acetate detection limit). There was very little difference in the effluent soluble COD concentrations between the two SBRs (Figure 3-52). With the exception of one CA data point, all values of effluent soluble COD were <9 mg/L.

3.5.2.3 End-of-Aerobic VSS and Sludge Storage Product Content

The total end-of-aerobic biomass levels of the AnA and CA SBRs were similar during the Loading Pattern 1 experiment and exhibited no significant trend (Figure 3-53). The measured AnA and CA end-of-aerobic VSS levels were typically within 10% of predicted values. The measured AnA end-of-aerobic PHA levels ranged from 10–15% of VSS (Figure 3-54), compared to a predicted value of approximately 12% of VSS. The measured CA end-of-aerobic PHA levels ranged between 7% and 9% of VSS, compared to a predicted value of 8.4% of VSS. The AnA and CA end-of-aerobic CH levels ranged from approximately 12–16% of VSS (predicted value of 14% of VSS) and 8% to 10% of VSS (predicted value of 9% of VSS) respectively (Figure 3-55). These results show that the sludge VSS and storage product contents were constant throughout the transient COD loading period for both sludge types, and that the AnA and CA experimental data were in excellent agreement with model predictions.

3.5.2.4 P Release and Acetate Uptake

For the AnA SBR, when the influent COD peak loading factor was 1.8 the P released during the anaerobic period was between 8.2 and 10.5 mgP/L (predicted 7.9–8.4 mgP/L), and when the influent COD peak loading factor was 0.2, the P released during the anaerobic period ranged between 0.9 to 1.6 mgP/L (predicted 2 mgP/L) (Figure 3-56). This adjustment in anaerobic P release was maintained throughout the 14-d experimental period. The measured values of P release were very close to the predicted values.

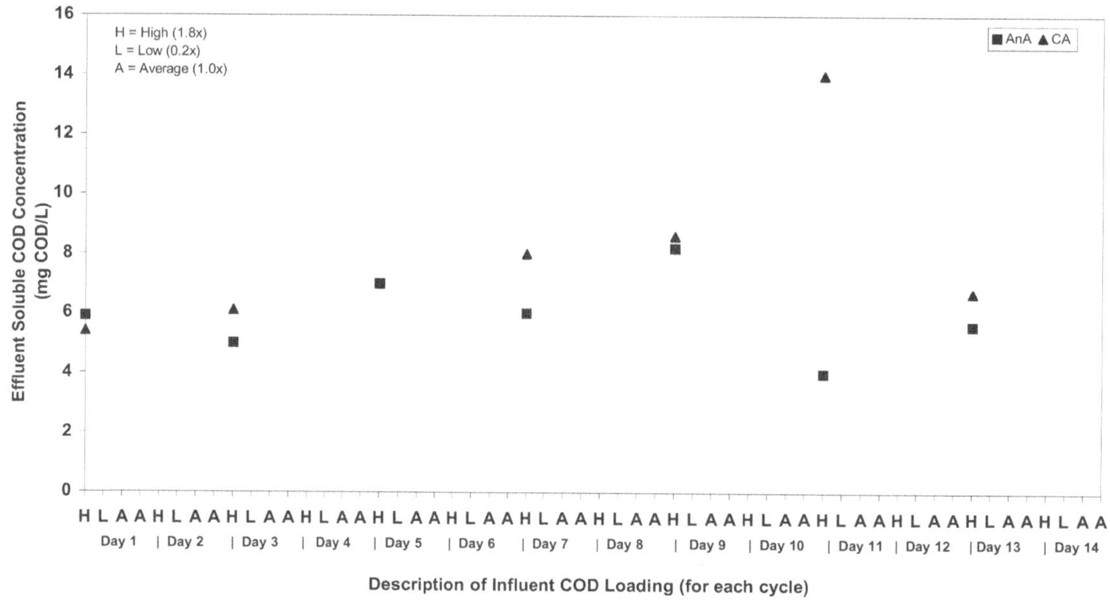

Figure 3-52. The Effect of Variable Influent COD Loading on Effluent Soluble COD Concentration of an AnA SBR and a CA SBR (Loading Pattern 1).

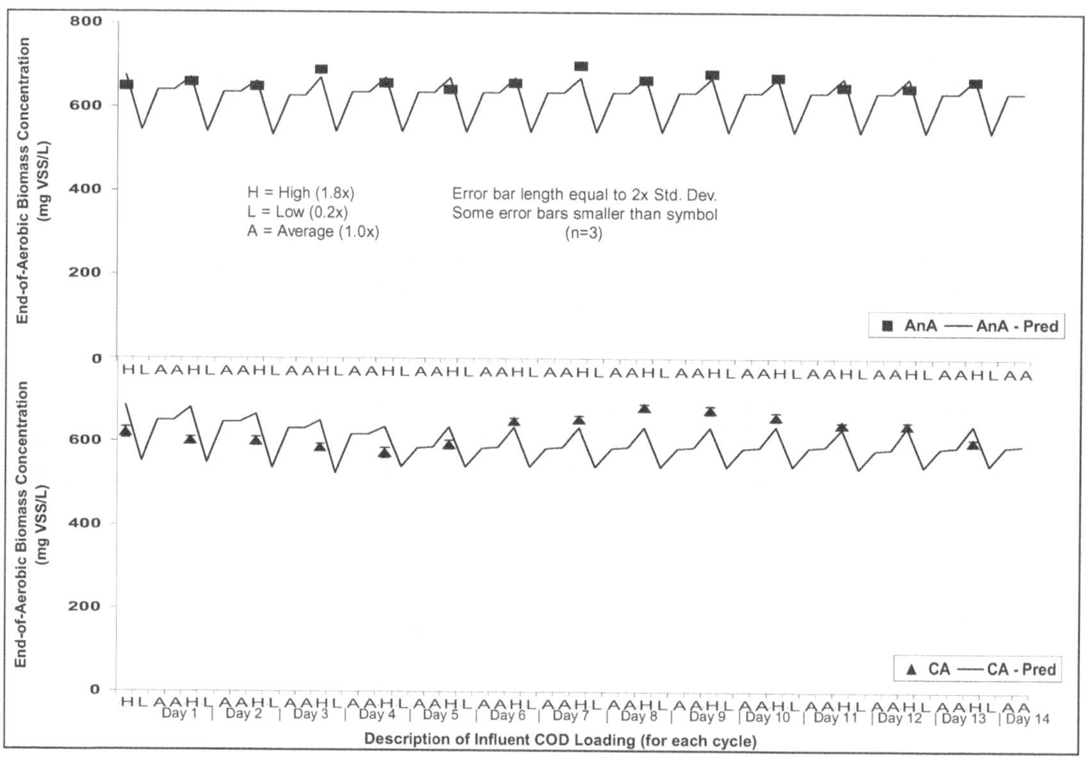

Figure 3-53. The Effect of Variable Influent COD Loading on End-of-Aerobic Biomass Concentration in an AnA SBR and a CA SBR (Loading Pattern 1).

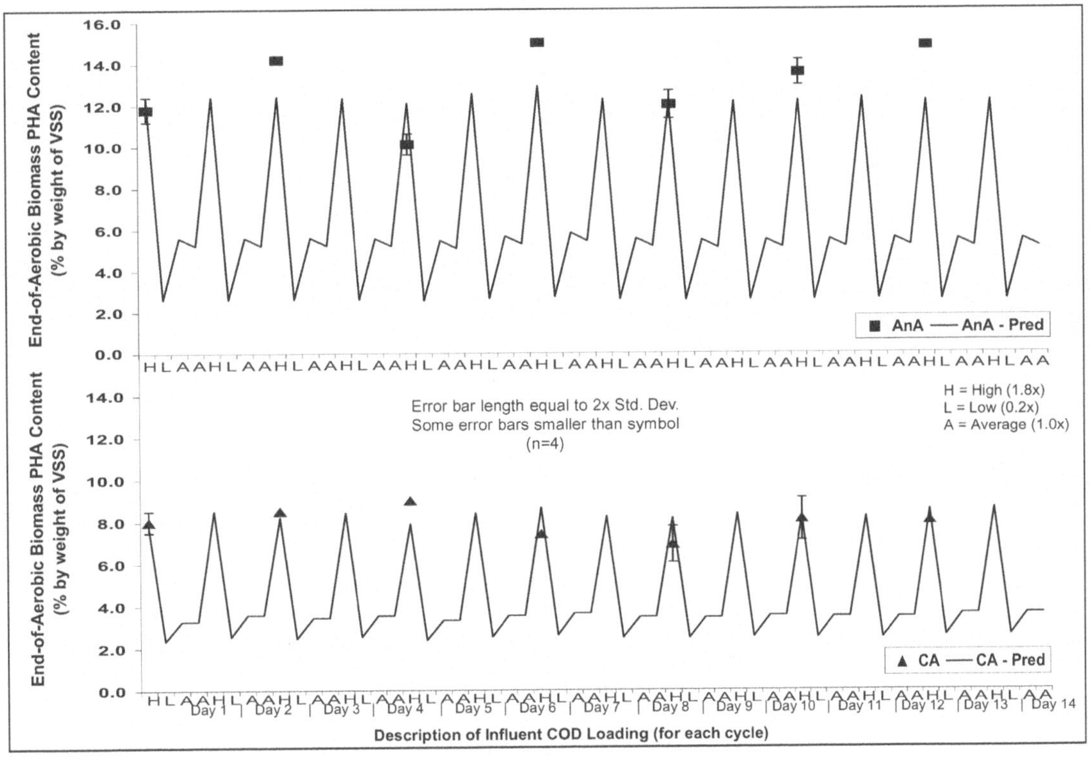

Figure 3-54. The Effect of Variable Influent COD Loading on End-of-Aerobic Biomass PHA Content in an AnA SBR and a CA SBR (Loading Pattern 1).

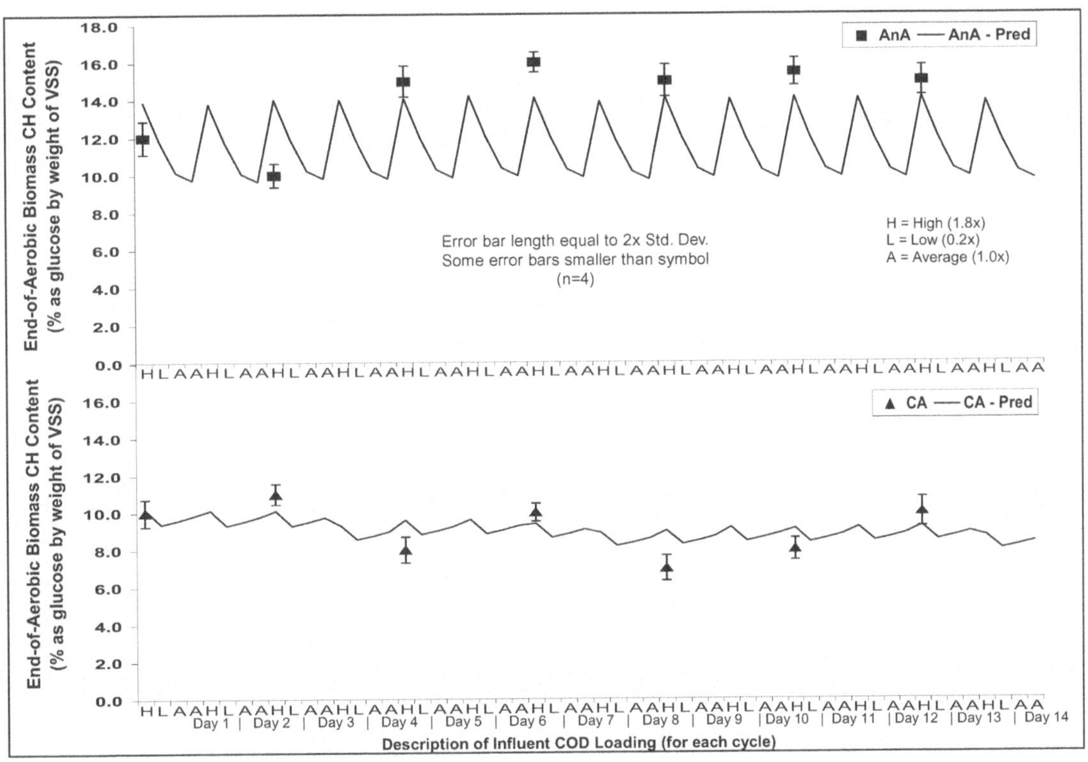

Figure 3-55. The Effect of Variable Influent COD Loading on End-of-Aerobic Biomass CH Content in an AnA SBR and a CA SBR (Loading Pattern 1).

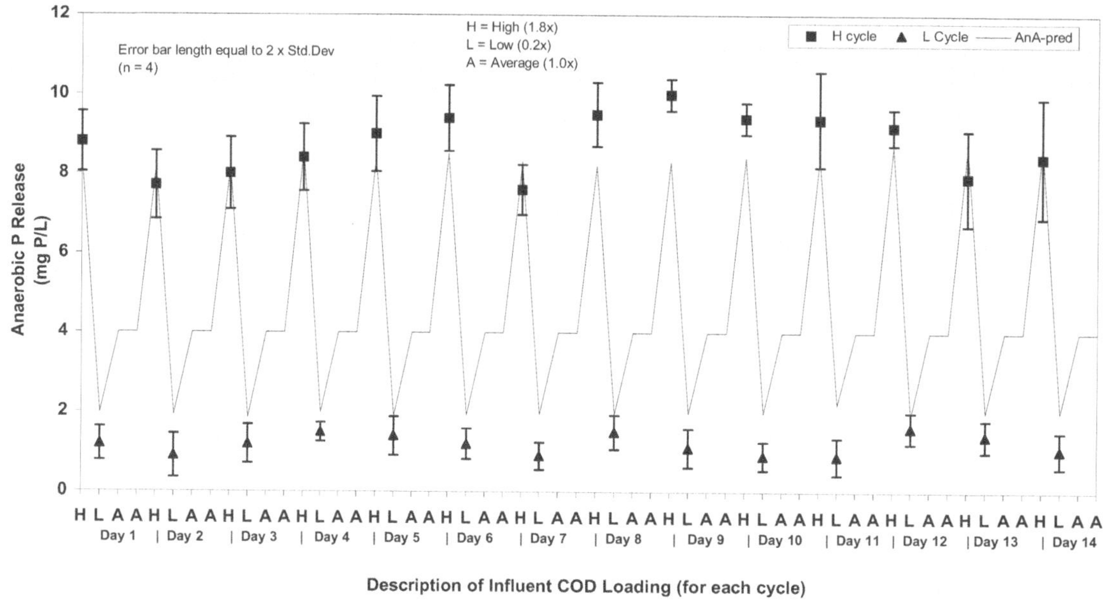

Figure 3-56. The Effect of Variable Influent COD Loading on P release from an AnA SBR (Loading Pattern 1).

The acetate uptake rate during the first 110 minutes of the high COD loading cycle was relatively stable for both sludge types (Figure 3-57). The measured AnA acetate uptake rate values were close to the predicted values (measured values of 0.66–0.9 mg acetic acid/L/min vs. predicted values of 0.78–0.9 mg acetic acid/L/min), as were those of the CA SBR (measured values of 1.15–1.45 mg acetic acid/L/min vs. predicted value of 1.3 mg acetic acid/L/min).

3.5.2.5 SVI and Effluent TSS

There was no significant difference in the SVI or the effluent TSS levels of the SBRs during the Loading Pattern 1 experiment (Figure 3-58).

3.5.3 Multiple Cycle Transient COD Loading Results: Loading Pattern 2, Low Weekend Loading

3.5.3.1 Effluent P Concentration

The AnA SBR maintained lower effluent soluble P concentrations than the CA SBR (Figure 3-59). The CA SBR effluent soluble P concentration averaged 1.3 mgP/L during weekend low COD loading periods (1.7 mgP/L predicted), and 0.8 mgP/L during weekday COD loading (0.9 mgP/L predicted). The AnA effluent soluble P concentrations averaged 0.4 mgP/L during weekend low COD loading periods (0.8 mgP/L predicted), and 0.3 mgP/L (0.3 mgP/L predicted) during weekday COD loading. Table 3-8 shows the daily average effluent soluble P concentrations for both SBRs.

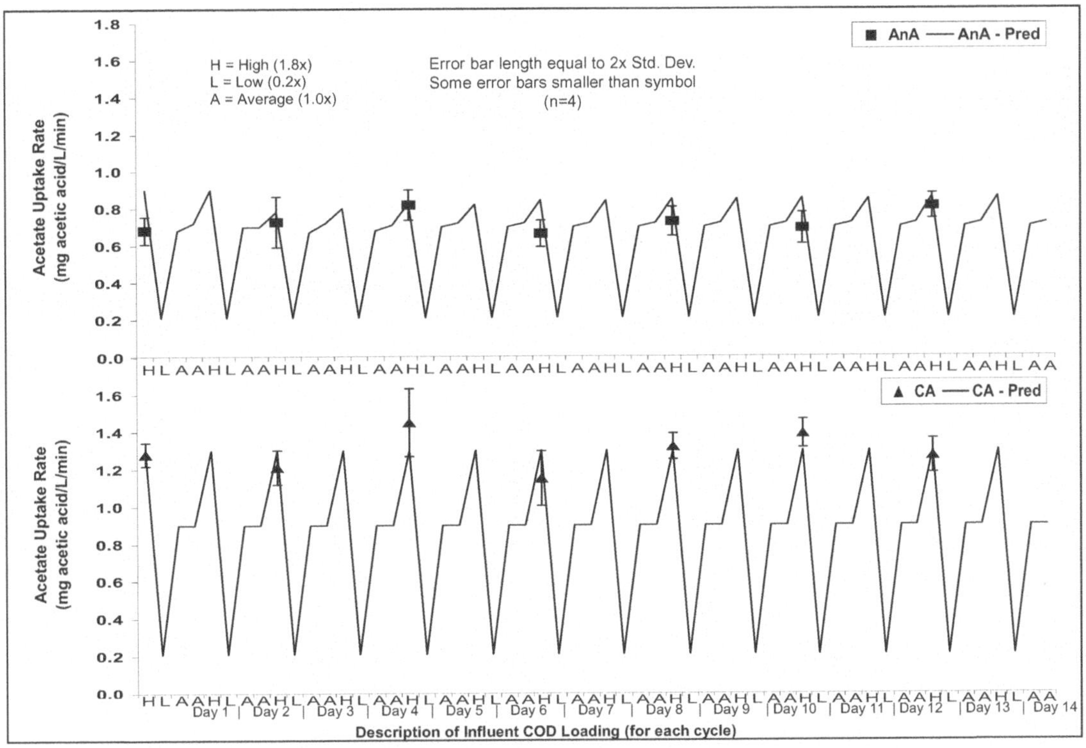

Figure 3-57. The Effect of Variable Influent COD Loading on Acetate Uptake Rate from an AnA SBR and a CA SBR (Loading Pattern 1).

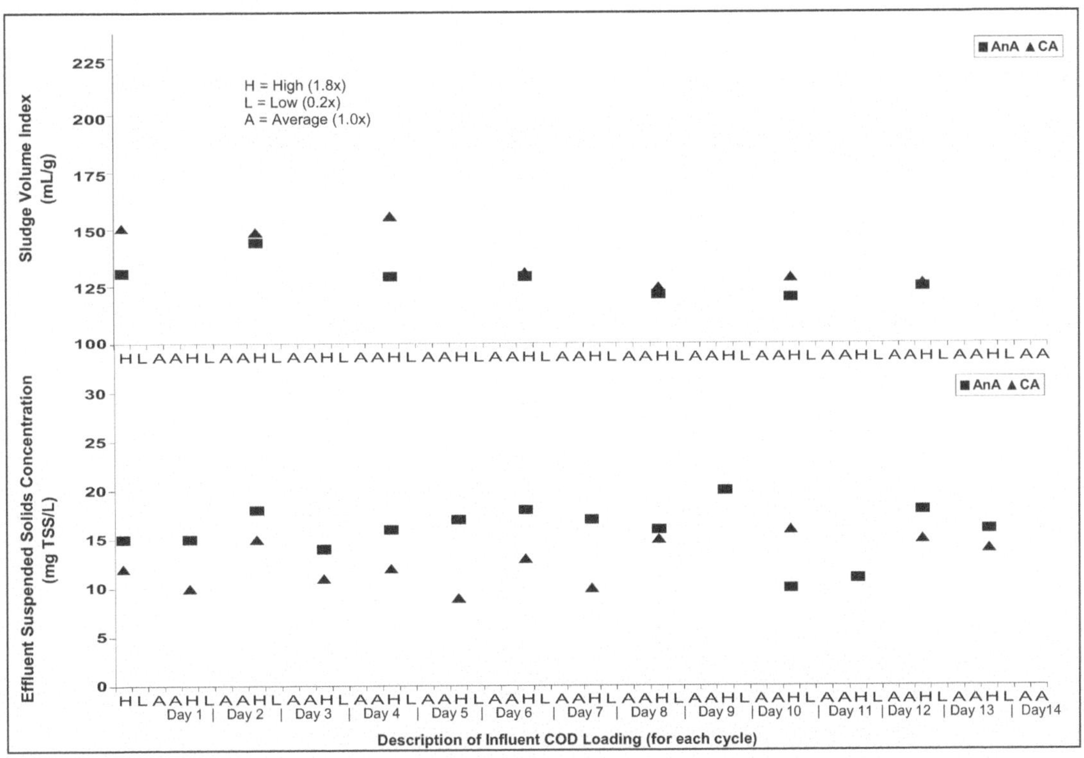

Figure 3-58. The Effect of Variable Influent COD Loading on Sludge Volume Index and Effluent Suspended Solids Concentration of an AnA SBR and a CA SBR (Loading Pattern 1).

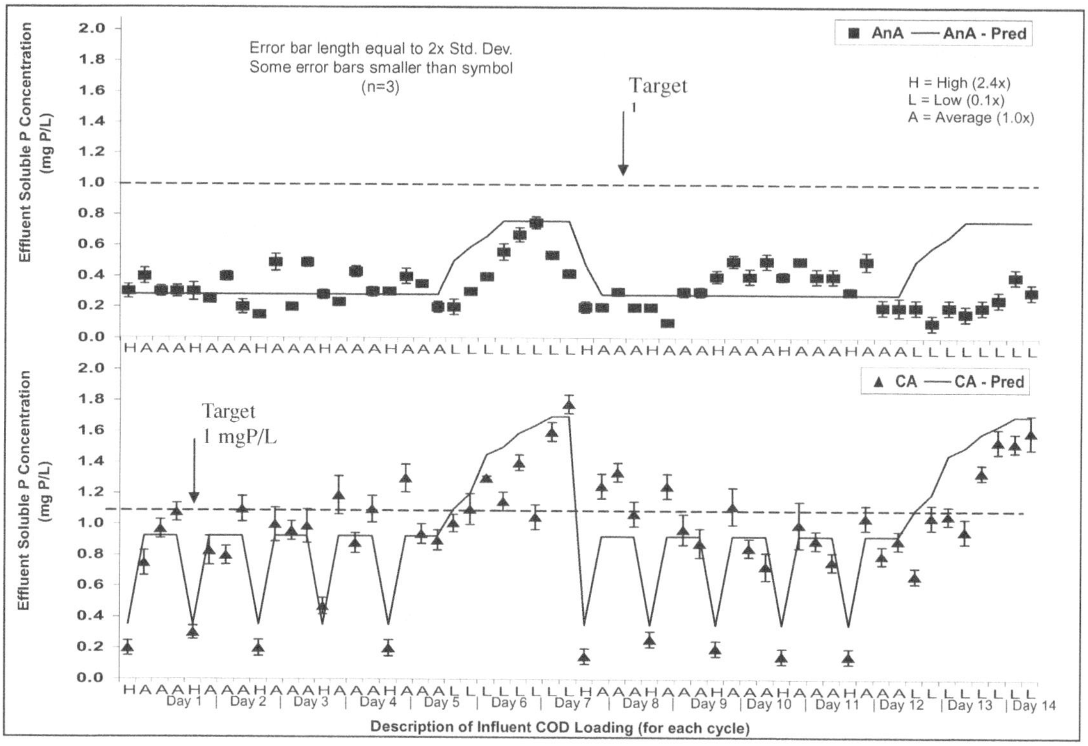

Figure 3-59. The Effect of Variable Influent COD Loading on Effluent Soluble P Concentration from an AnA SBR and a CA SBR (Loading Pattern 2).

Table 3-8. Daily Average Effluent Soluble P Concentration Loading Pattern 2 Experiment.

Day (Pattern)	AnA, mg P/L	CA, mg P/L
1 (H-A-A-A)	0.3	0.8
2 (H-A-A-A)	0.3	0.8
3 (H-A-A-A)	0.3	0.8
4 (H-A-A-A)	0.3	0.9
5 (H-A-A-A)	0.3	0.8
6 (L-L-L-L)	0.4	1.1
7 (L-L-L-L)	0.6	1.5
8 (H-A-A-A)	0.2	1.0
9 (H-A-A-A)	0.2	0.8
10 (H-A-A-A)	0.5	0.7
11 (H-A-A-A)	0.4	0.7
12 (H-A-A-A)	0.3	0.7
13 (L-L-L-L)	0.2	0.9
14 (L-L-L-L)	0.3	1.5
Overall Averages	0.3	0.9
Loading Pattern Symbols: H = 2.4x, A = 1.0x, L = 0.1x		

3.5.3.2 Effluent Acetate and Soluble COD Concentration

Effluent acetate and soluble COD samples were taken after the high COD loading cycle for each SBR (Figures 3-60 and 3-61). Acetate was detected in the effluent stream of the AnA SBR (15 mg/L) and of the CA SBR (13 mg/L) following the first high COD loading cycle after the low COD loading period. The predicted effluent acetate concentrations for both SBRs over-estimated the measured values by approximately 5 mg/L. The effluent soluble COD concentrations for the CA and AnA SBRs were similar and ranged between 5 and 10 mg/L when no acetate was detected in the effluent. These results show that acetate breakthrough occurred for both the CA and AnA SBR after the first high COD loading cycle following the weekend low COD loading period.

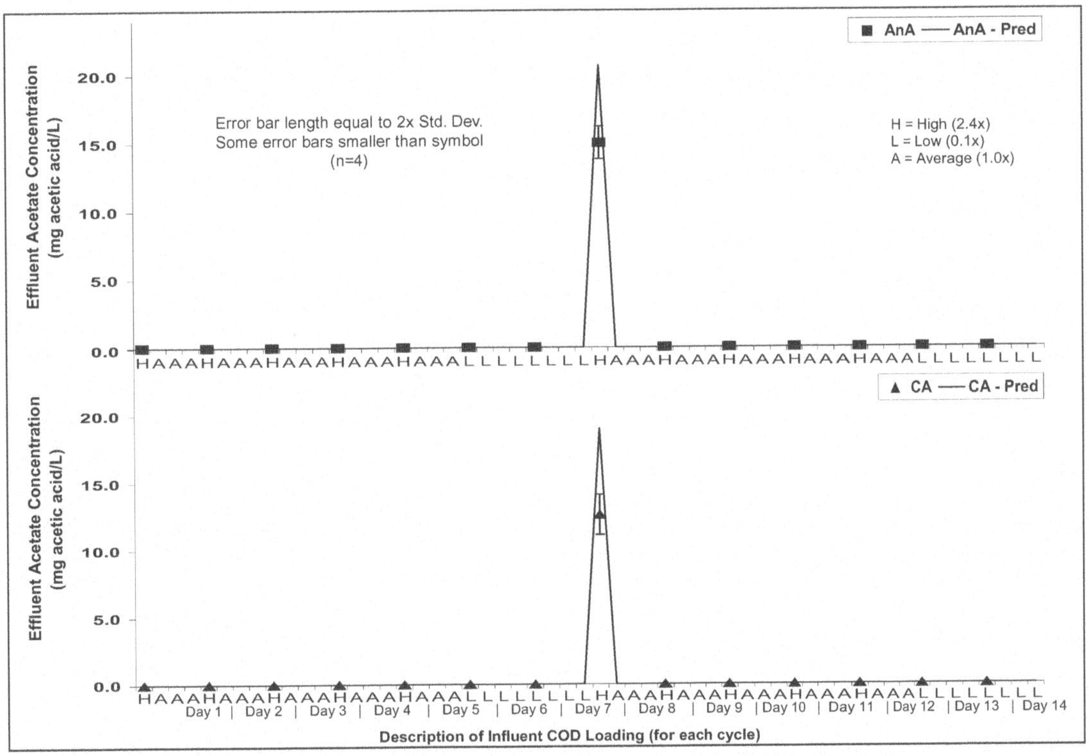

Figure 3-60. The Effect of Variable Influent COD Loading on Effluent Acetate Concentration from an AnA SBR and a CA SBR (Loading Pattern 2).

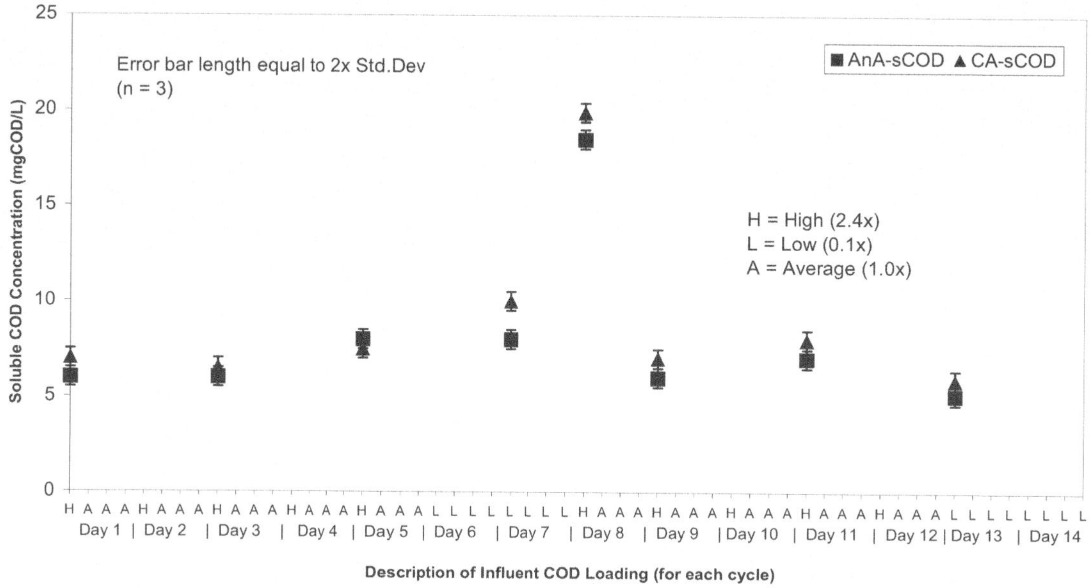

Figure 3-61. The Effect of Variable Influent COD Loading of Effluent Soluble COD Concentration of an AnA SBR and a CA SBR (Loading Pattern 2).

3.5.3.3 End-of-Aerobic VSS and Sludge Storage Product Content

The end-of-aerobic biomass concentrations in the AnA and CA SBRs were similar throughout the experiment (Figure 3-62). The end-of-aerobic VSS levels dropped during the weekend low COD loading period but increased when weekday COD loading conditions resumed. Note that there is a gradual downward trend in the VSS of both reactors throughout the 14-d experiment. This probably indicates that the long-term biomass content of both SBRs is less than the steady state biomass content under uniform COD loading conditions (long-term biomass stability is further investigated in Section 3.5.5.2.). The measured AnA and CA end-of-aerobic VSS values were usually within 10% of the predicted values. The AnA sludge end-of-aerobic PHA level also decreased to approximately 2% of VSS during the weekend low COD loading period but then increased to approximately 13–16% of VSS when weekday COD loading resumed (Figure 3-63). The predicted AnA sludge end-of-aerobic PHA values were 14% of VSS during weekday COD loading, and 2.4% of VSS during weekend COD loading. The CA sludge maintained a PHA content of approximately 4–6% of VSS throughout the experiment. The predicted values ranged from 3.5–5% of VSS. The end-of-aerobic CH contents of both sludge types were between 13% and 19% (Figure 3-64). The predicted end-of-aerobic CH contents of both the AnA and CA sludge were 13–16% and 13–15% of VSS, respectively.

3.5.3.4 P Release and Acetate Uptake

The anaerobic P release in the AnA SBR was typically approximately 7.0 mg P/L during high influent COD loading conditions, and from 0.3–0.5 mg P/L during low COD loading conditions (Figure 3-65). On Day 8, the mass of P released during the high COD loading cycle was approximately 30% smaller than it was on Day 5, prior to the low weekend COD loading period. This reduction in anaerobic P release may indicate that the PAM sequestered less acetate during the first high loading cycle following the weekend. The mass of P release increased to approximately 7 mgP/L on Days 9 through 12. The measured values of P release were generally close to predicted values.

The concentration of acetate removed during the first 110 minutes of each SBR cycle was relatively constant, and was higher for the CA SBR than for the AnA SBR (Figure 3-66). The measured AnA and CA acetate uptake agreed well with predicted data.

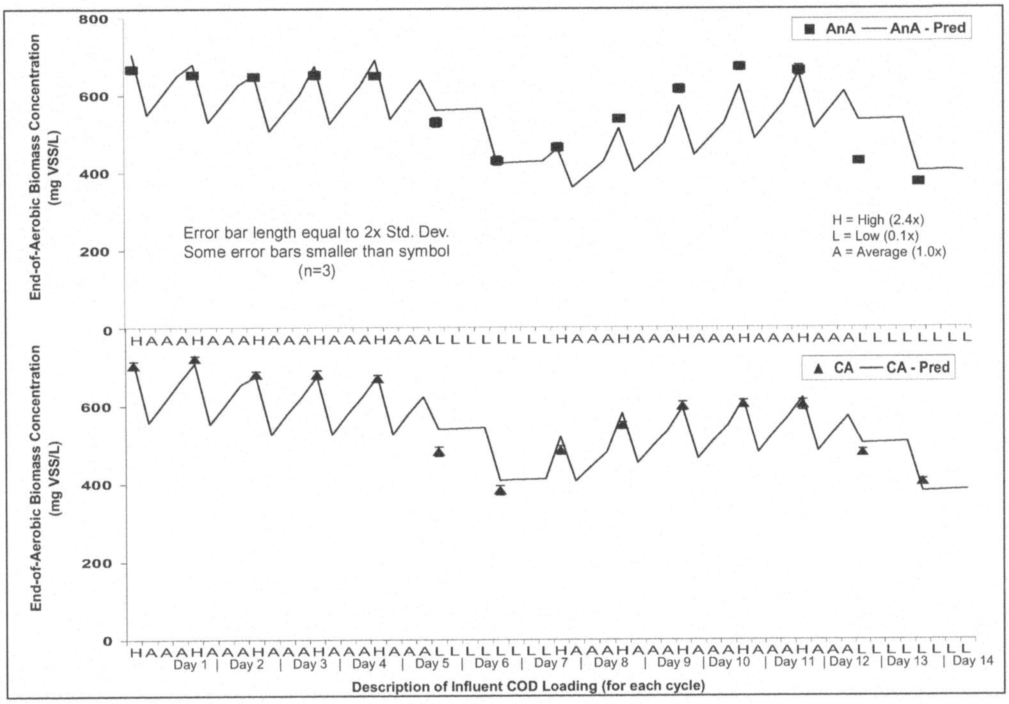

Figure 3-62. The Effect of Variable Influent COD Loading on End-of-Aerobic Biomass Concentration from an AnA SBR and a CA SBR (Loading Pattern 2).

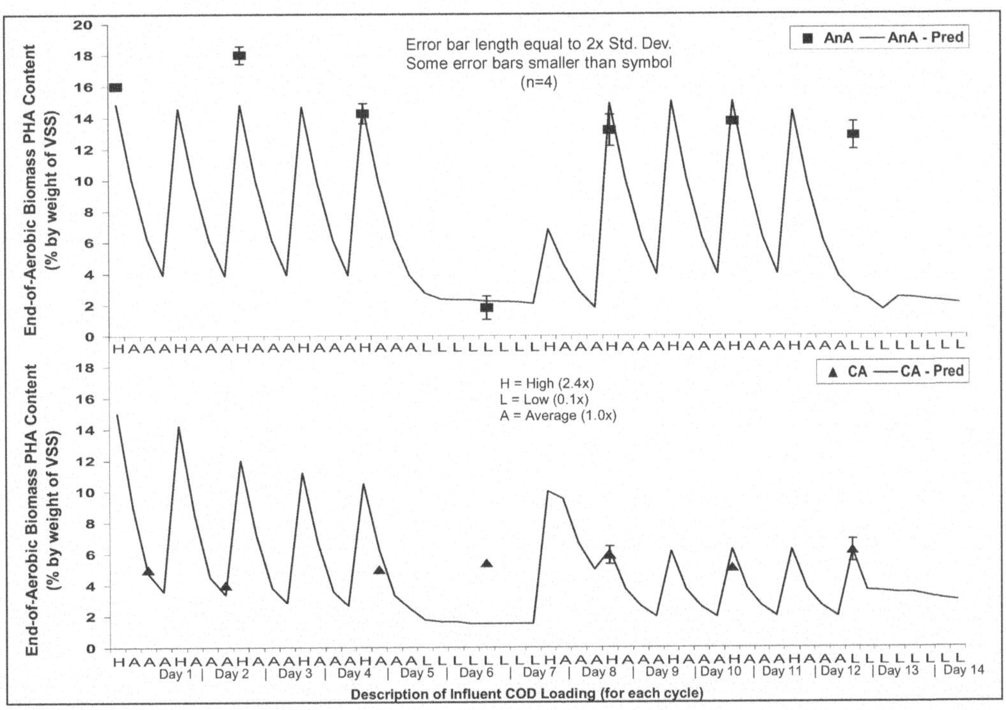

Figure 3-63. The Effect of Variable Influent COD Loading on End-of-Aerobic Biomass PHA Content from an AnA SBR and a CA SBR (Loading Pattern 2).

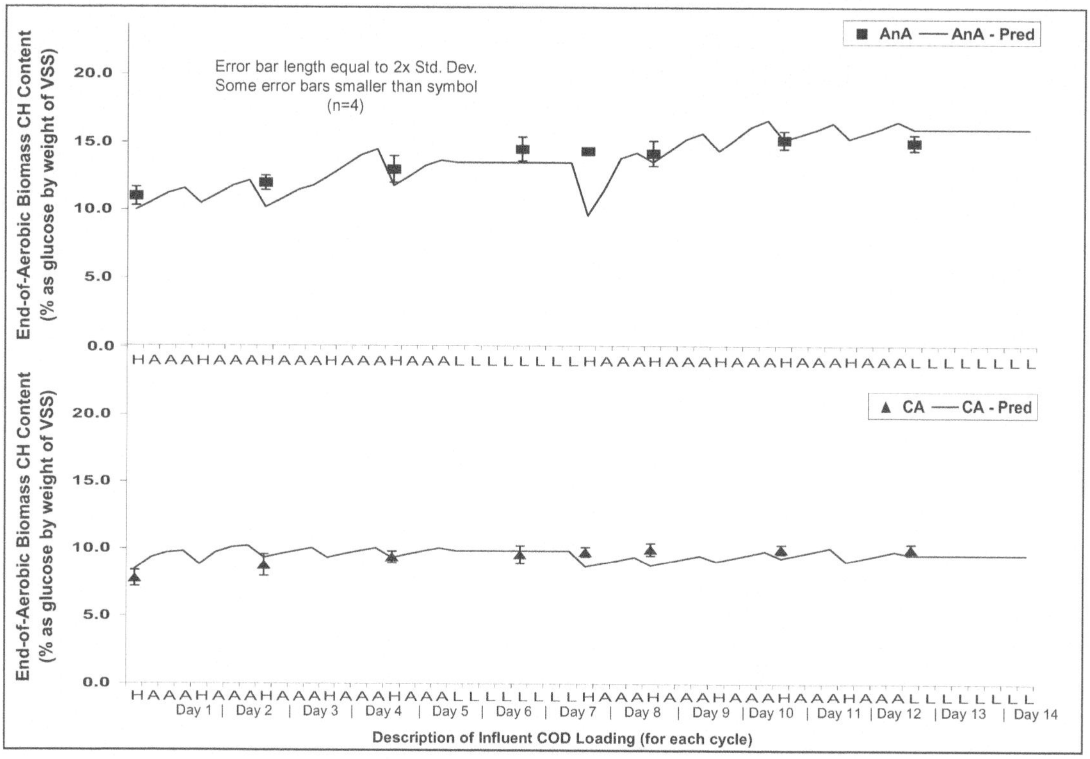

Figure 3-64. The Effect of Variable Influent COD Loading on End-of-Aerobic Biomass CH Content from an AnA SBR and a CA SBR (Loading Pattern 2).

3.5.3.5 SVI and Effluent TSS

The SVI of both SBRs increased during the low COD loading period due to changes in VSS concentration, but the effluent TSS values of both SBRs were ≤20 mg/L throughout the experiment (Figure 3-67).

3.5.3.6 Process Failure

During the Loading Pattern 2 experiment, process failure due to the presence of effluent acetate was observed during the first high COD loading cycle following the low COD loading weekend period (Figure 3-61, Day 8). This failure occurred because the VSS levels in the reactors had fallen from 650 to 430 mg/L (34% reduction) for the AnA SBR and from 675 to 383 mg/L (43% reduction) for the CA SBR during the two days of low COD loading. The results of the single cycle transient COD loading experiments showed that when the influent COD peak loading factor was ≥2.6, effluent acetate was detected for the AnA and CA SBRs (Figure 3-41). The reduction in VSS levels during the Loading Pattern 2 experiment allowed acetate breakthrough at an influent COD peak loading factor of 2.4.

The experimental data also point toward another possible mode of process failure. On Day 8 of the Loading Pattern 2 experiment, the mass of P release during the high COD loading cycle was approximately 30% smaller than it was on Day 5, prior to the low weekend COD loading period (Figure 3-65). This reduction in anaerobic P release may indicate that the PAM sequestered less acetate during the first high COD loading cycle following the weekend. This could lead to a gradual reduction in the PAM fraction of the AnA VSS, and process failure due to elevated soluble P concentrations (this is investigated in Section 3.5.5.2.).

3.5.4 Multiple Cycle Transient COD Loading Results: Loading Pattern 3, Extended Low Loading

3.5.4.1 Effluent P Concentration

The average effluent soluble P concentration of both the SBRs was ≥1 mgP/L during this experiment (Figure 3-68). The overall average effluent soluble P concentrations from the AnA and CA SBR were 1.1 and 1.5 mgP/L, respectively (Table 3-9). During the extended low COD loading period, the average effluent soluble P concentrations from the AnA and CA SBRs were 1.1 and 1.7 mgP/L, respectively. The measured values of AnA and CA effluent soluble P concentrations followed the same trend as predicted values.

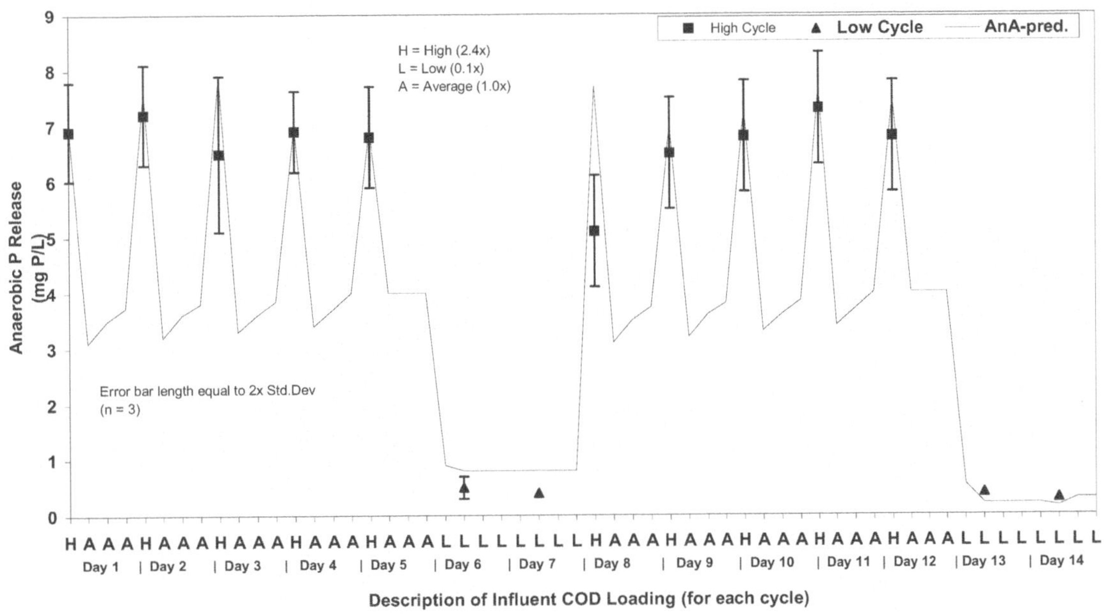

Figure 3-65. The Effect of Variable Influent COD Loading on P release from an AnA SBR (Loading Pattern 2).

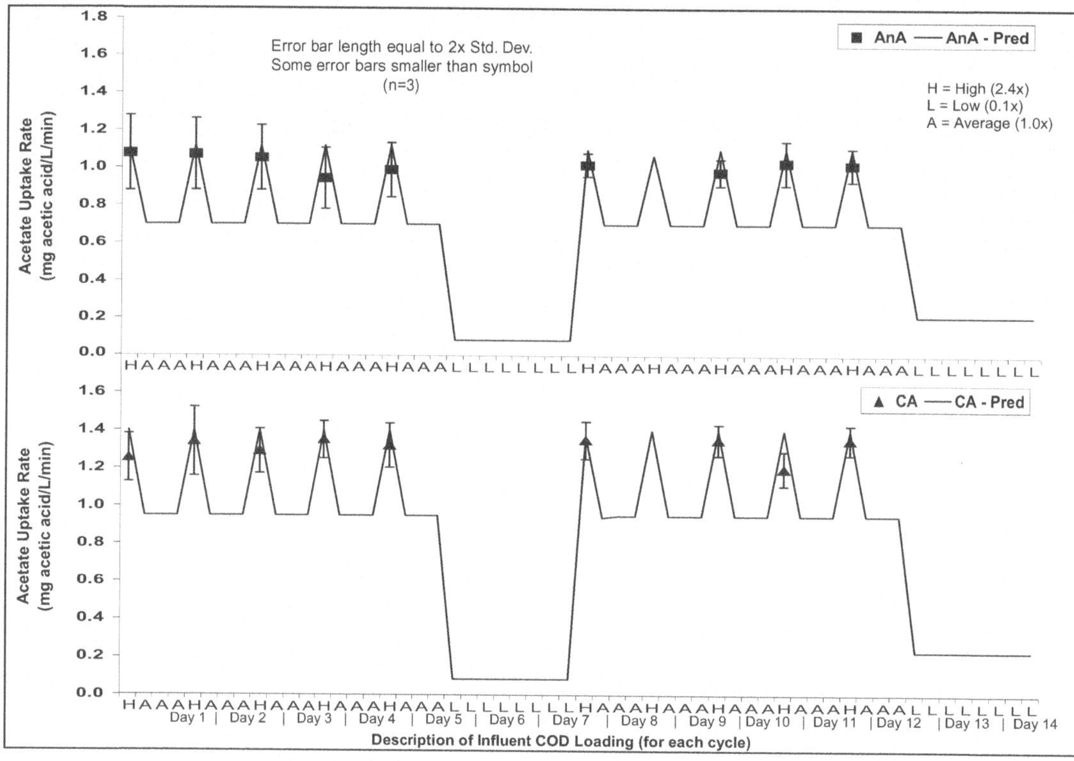

Figure 3-66. The Effect of Variable Influent COD Loading on Acetate Uptake Rate from an AnA SBR and a CA SBR (Loading Pattern 2).

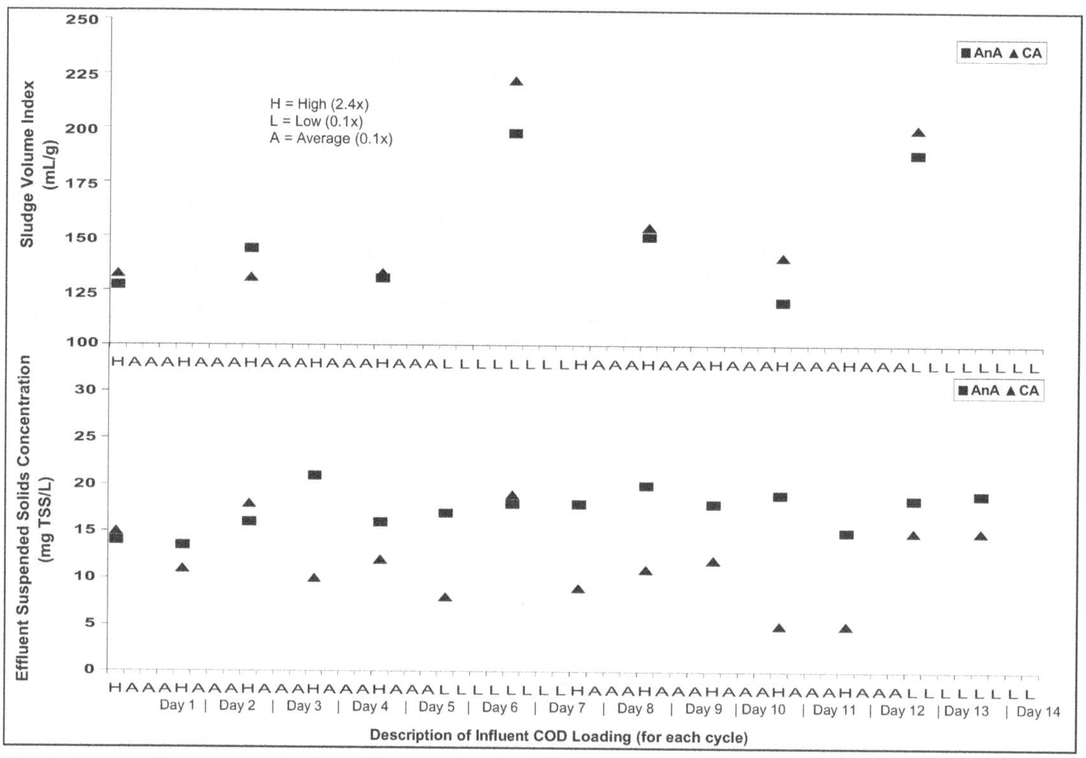

Figure 3-67. The Effect of Variable Influent COD Loading on Sludge Volume Index and Effluent Suspended Solids Concentration an AnA SBR and a CA SBR (Loading Pattern 2).

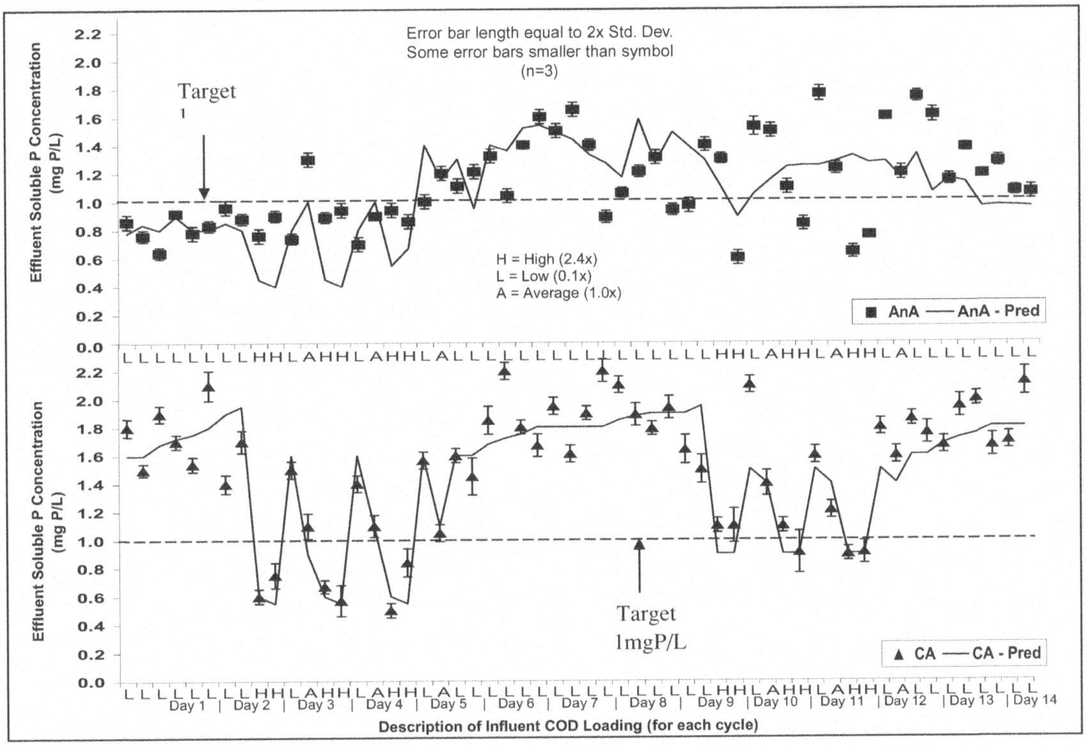

Figure 3-68. The Effect of Variable Influent COD Loading on Effluent Soluble P Concentration from an AnA SBR and a CA SBR (Loading Pattern 3).

Table 3-9. Daily Average Effluent Soluble P Concentration Loading Pattern 3 Experiment.

Day (Pattern)	AnA, mg P/L	CA, mg P/L
1 (L-L-L-L)	0.8	1.7
2 (L-L-L-L)	0.9	1.7
3 (H-H-L-A)	0.9	1.0
4 (H-H-L-A)	0.9	0.9
5 (H-H-L-A)	1.0	1.0
6 (L-L-L-L)	1.2	1.8
7 (L-L-L-L)	1.5	1.8
8 (L-L-L-L)	1.1	2.0
9 (L-L-L-L)	1.2	1.7
10 (H-H-L-A)	1.2	1.4
11 (H-H-L-A)	1.2	1.2
12 (H-H-L-A)	1.1	1.3
13 (L-L-L-L)	1.5	1.8
14 (L-L-L-L)	1.1	1.9
Overall Averages	1.1	1.5
Loading Pattern Symbols: H = 2.4x, A = 1.0x, L = 0.1x		

3.5.4.2 Effluent Acetate and Soluble COD Concentration

Elevated effluent soluble COD and acetate concentrations were detected for both SBRs (Figures 3-69 and 3-70). Acetate was detected in five AnA and CA effluent samples, each averaging 8 mg/L in both instances. The model predicted the appearance of acetate in the same cycles during which it was detected experimentally for both SBRs. The predicted effluent acetate concentrations are within 15% of the measured values.

3.5.4.3 End-of-Aerobic VSS and Sludge Storage Product Content

The end-of-aerobic biomass contents of both SBRs dropped sharply during the extended low COD loading period (long weekend); and recovered only partially when the weekday COD loading conditions resumed (Figure 3-71). The measured AnA and CA end-of-aerobic VSS values follow the predicted trend. The end-of-aerobic PHA contents of both sludges were between 2–6% of VSS during the extended (4-d weekend) low COD loading period, and between 8–18% of VSS during the weekday COD loading conditions (Figure 3-72). The measured AnA PHA contents corresponded well with predicted values through Day 8, but for the AnA PHA values on Days 11 and 13, the AnA model underpredicted the measured values by approximately 15%. The measured values followed the CA model predictions closely. Over the experimental period the AnA end-of-aerobic CH contents dropped from approximately 14–6% of VSS (predicted values dropped from 14–9% of VSS) (Figure 3-73). The CA end-of-aerobic CH contents ranged between 4% and 7% of VSS (with one exception). The predicted CA end-of-aerobic CH content ranged between 5% and 7% of VSS.

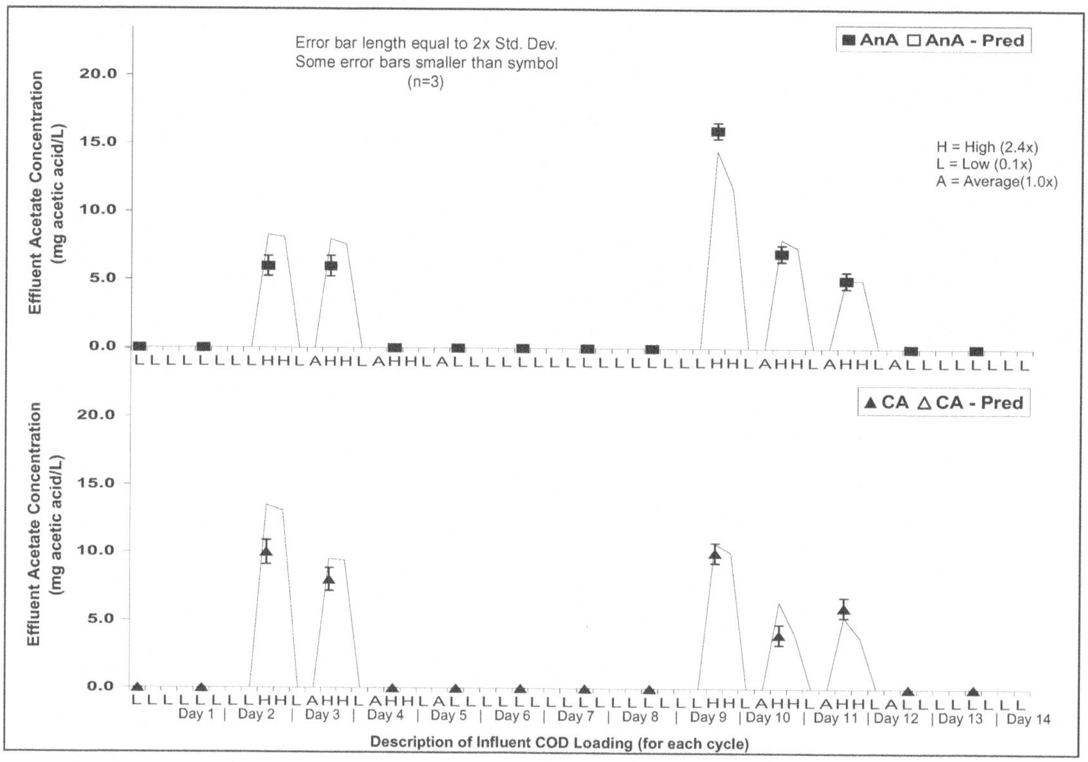

Figure 3-69. The Effect of Variable Influent COD Loading on Effluent Acetate Concentration from an AnA SBR and a CA SBR (Loading Pattern 3).

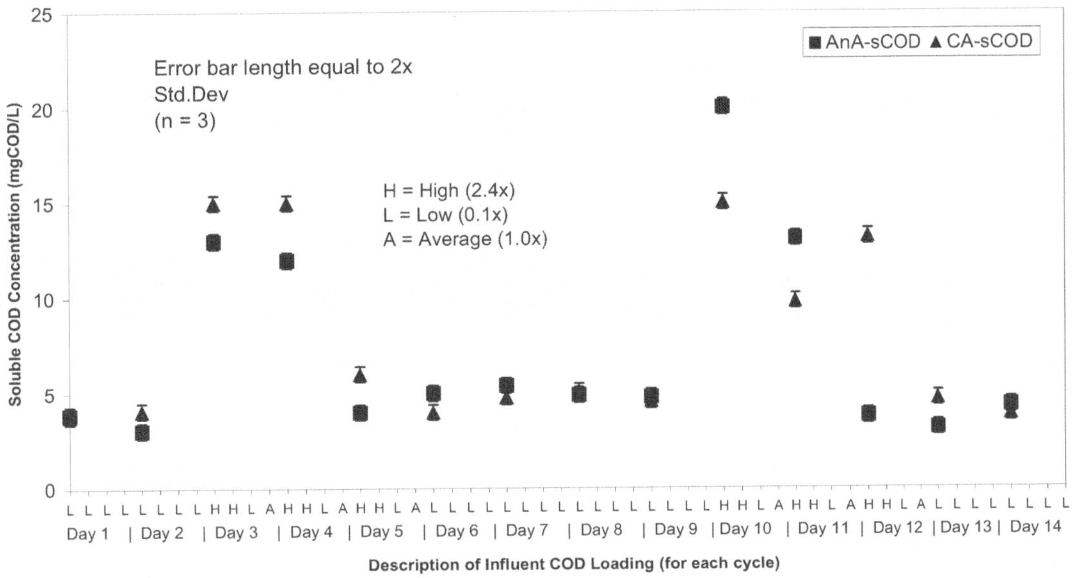

Figure 3-70. The Effect of Variable Influent COD Loading of Effluent Soluble COD Concentration of an AnA SBR and a CA SBR (Loading Pattern 3).

3.5.4.4 P Release and Acetate Uptake

The anaerobic P release in the AnA SBR decreased to zero during the 16-cycle low COD loading period (4-d weekend), and only increased to 3 mgP/L (about half of its original value) when the normal "weekday" COD loading resumed (Figure 3-74). The observed decrease in P release is likely due to the drop in total VSS (Figure 3-71). The AnA P release data corresponded with predicted values during normal "weekday" COD loading conditions, but were lower than the predicted values by approximately 0.5 mgP/L during the low COD loading period. Acetate uptake rates observed during the second week were slightly lower than those measured during the first week (Figure 3-75), partially due to the reduction in end-of-aerobic VSS that occurred as a result of the extended low COD loading period. The experimentally determined trend followed model predictions.

3.5.4.5 SVI and Effluent TSS

The SVI and effluent TSS concentrations of both SBRs increased through Day 8 (Figure 3-76). All SVI values were ≥150 mL/g, and the effluent TSS values reached 40 mg/L. This result showed that Loading Pattern 3 resulted in a decrease in TSS removal efficiency in both reactors.

3.5.4.6 Process Failure

During the Loading Pattern 3 experiment, both the AnA and CA process failed due to the presence of elevated effluent acetate and soluble P concentrations (Figure 3-69, 3-70). The 4-d low COD loading period caused a sharp reduction in the reactor VSS levels (Figure 3-71), which reduced the capacity of the both SBRs to remove influent COD.

Elevated soluble P concentrations were observed in the AnA effluent during the low COD loading period of the Loading Pattern 3 experiment, particularly during and after the extended low COD loading period (Figure 3-69). There was very little anaerobic P release during the extended low COD loading period, and when "normal" COD loading resumed, only 2–3 mgP/L was anaerobically released (Figure 3-74). These observations suggested that the extended low COD loading period reduced the PAM activity of AnA sludge.

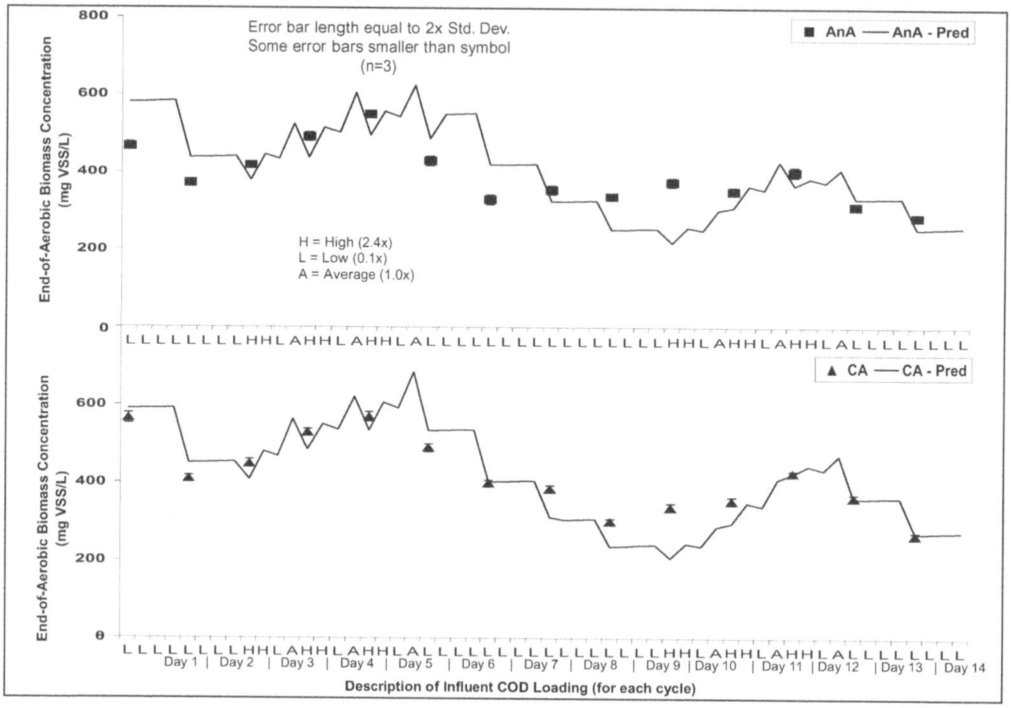

Figure 3-71. The Effect of Variable Influent COD Loading on End-of-Aerobic Biomass Concentration from an AnA SBR and a CA SBR (Loading Pattern 3).

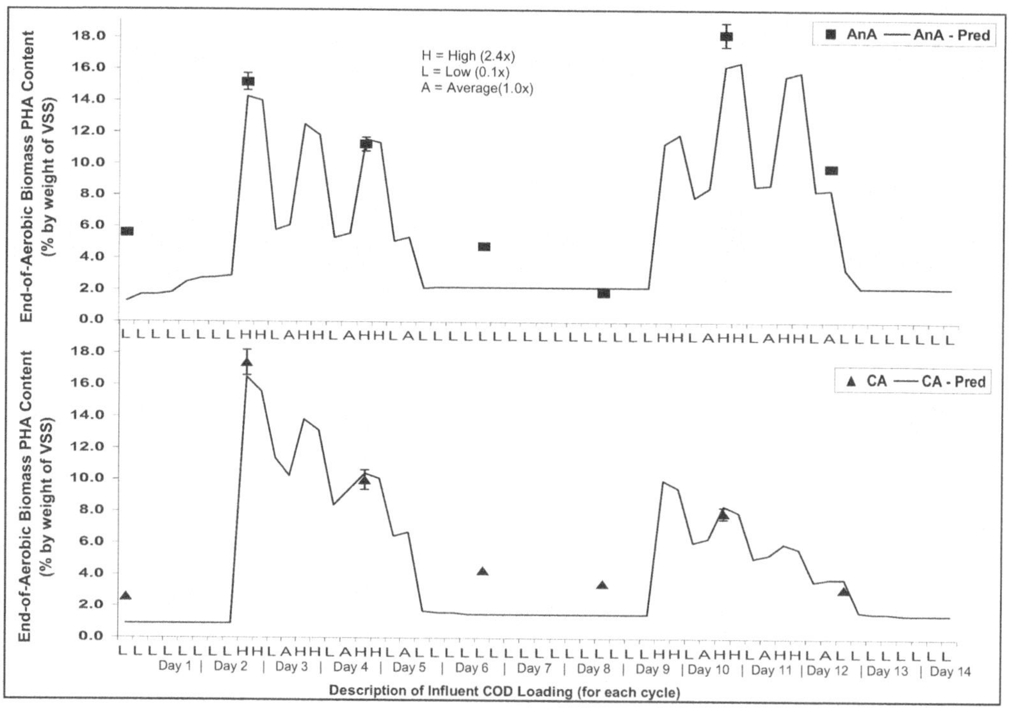

Figure 3-72. The Effect of Variable Influent COD Loading on End-of-Aerobic Biomass PHA Content from an AnA SBR and a CA SBR (Loading Pattern 3).

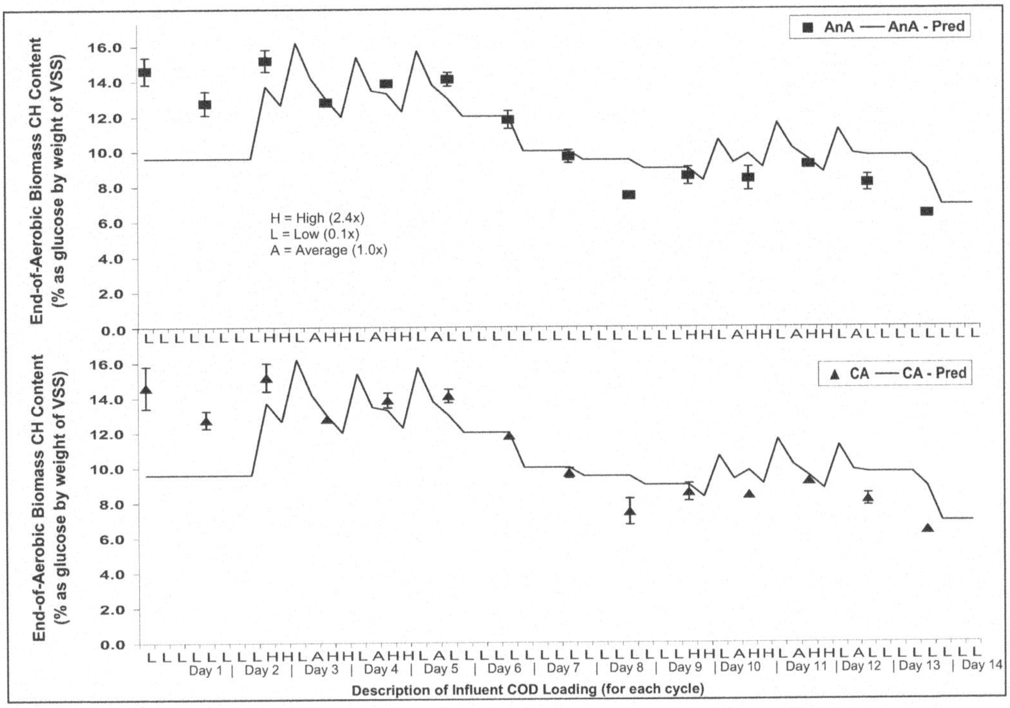

Figure 3-73. The Effect of Variable Influent COD Loading on End-of-Aerobic Biomass CH Content from an AnA SBR and a CA SBR (Loading Pattern 3).

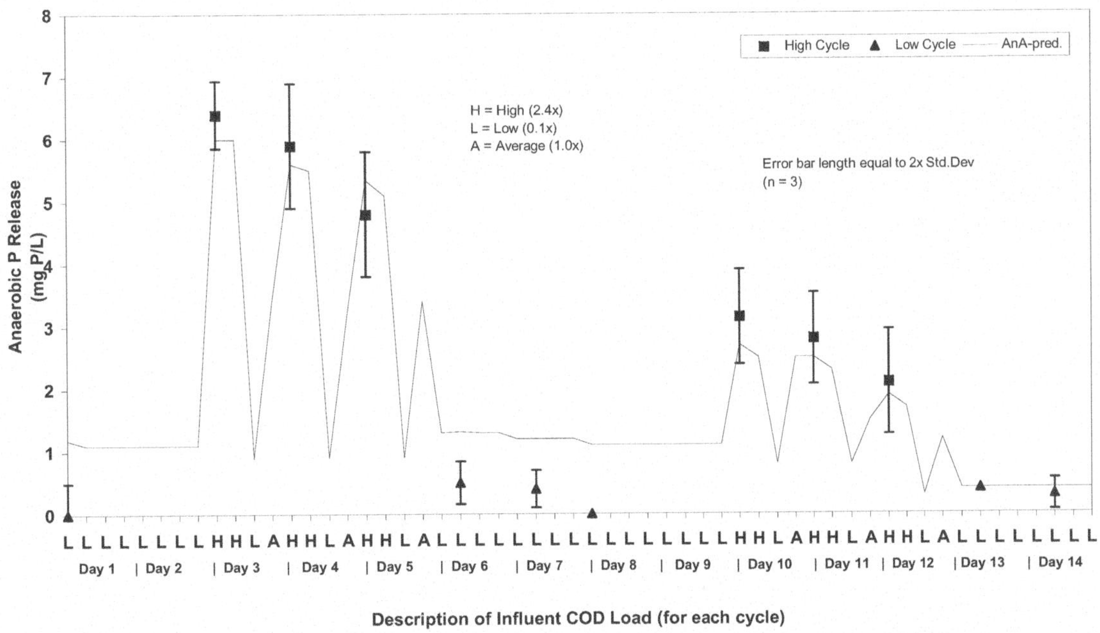

Figure 3-74. The Effect of Variable Influent COD Load on P Release from an AnA SBR (Loading Pattern 3).

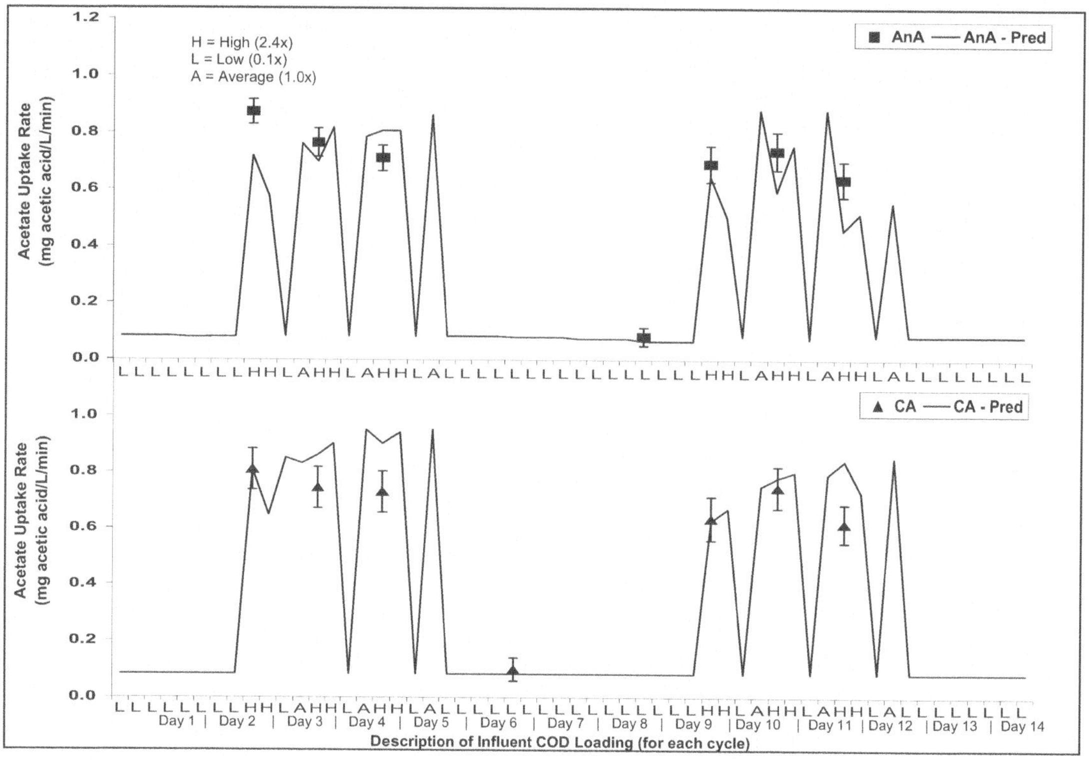

Figure 3-75. The Effect of Variable Influent COD Loading on Acetate Uptake Rate from an AnA SBR and a CA SBR (Loading Pattern 3).

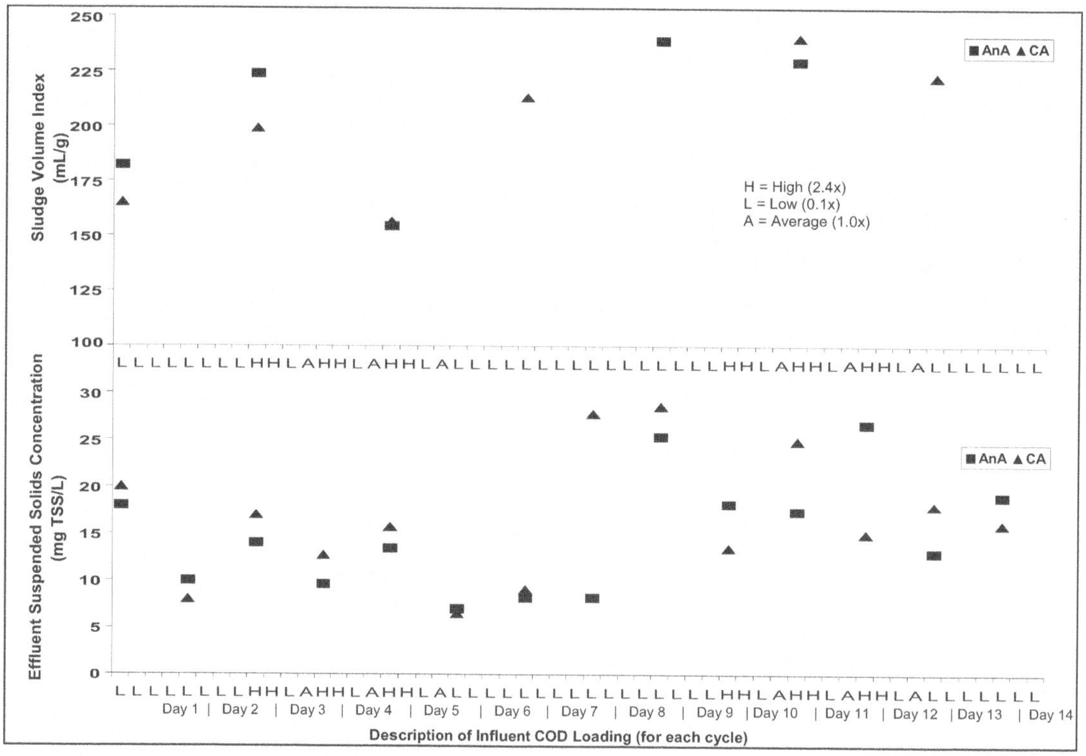

Figure 3-76. The Effect of Variable Influent COD Loading on Sludge Volume Index and Effluent Suspended Solids Concentration an AnA SBR and a CA SBR (Loading Pattern 3).

3.5.5 Long-Term Simulations

Simulations were conducted to investigate long-term process stability and modes of failure when the AnA and CA systems were subjected to Loading Patterns 1 through 3 over an 800-d period.

3.5.5.1 Loading Pattern 1

Model predictions for AnA operation show that the total biomass, PAM and GAM biomass fractions (Figure 3-77), end-of-aerobic PHA and CH content (Figure 3-78) remained stable during the 800-d simulation when subjected to Loading Pattern 1. Anaerobic P release persisted (Figure 3-79). Effluent acetate and P concentrations averaged 0 and 0.3 mg P/L (Figure 3-80). CA simulations showed a stable cyclical variation in biomass and storage product levels (Figures 3-81 and 3-82). The effluent acetate and soluble P concentrations averaged 0 and 1.0 mgP/L (Figure 3-83). PAM or GAM washout did not occur in the AnA system, and biomass washout did not occur in the CA system. Both systems operated in stable fashion over the 800-d period.

3.5.5.2 Loading Pattern 2

During the Loading Pattern 2 simulation, the model predictions for AnA operation show that the total biomass concentration remained stable, but that the PAM biomass fraction gradually decreased from 0.15 to 0.02 (Figure 3-84). This led to elevated effluent soluble P concentrations during the low COD loading period (Figure 3-85) in addition to weekly breakthrough of effluent acetate (after the first high loading cycle following the weekend low COD loading period). The predicted anaerobic P release profile showed that the amount of P release during the first high COD loading cycle of the week (following the low weekend COD loading period) gradually decreased because the PAM gradually captured less acetate during the first high loading cycle of the week (Figure 3-86). Thus, the AnA model simulation suggested that a long-term PAM-related process failure may occur when an influent COD loading pattern includes a low weekend COD loading period. The AnA end-of-aerobic PHA and CH contents were stable (Figure 3-87). The CA model predictions showed that biomass and storage product levels remained stable (Figures 3-88 and 3-89), but that elevated effluent soluble P concentrations and weekly breakthrough of acetate occurred following the first high loading cycle after the low COD loading period (Figure 3-90).

3.5.5.3 Loading Pattern 3

For both the AnA and CA systems, the total biomass content decreased and elevated effluent acetate and P concentrations were predicted (Figures 3-91, 3-92, 3-93, 3-94), consistent with the results and discussion presented in Section 3.5.4.6.

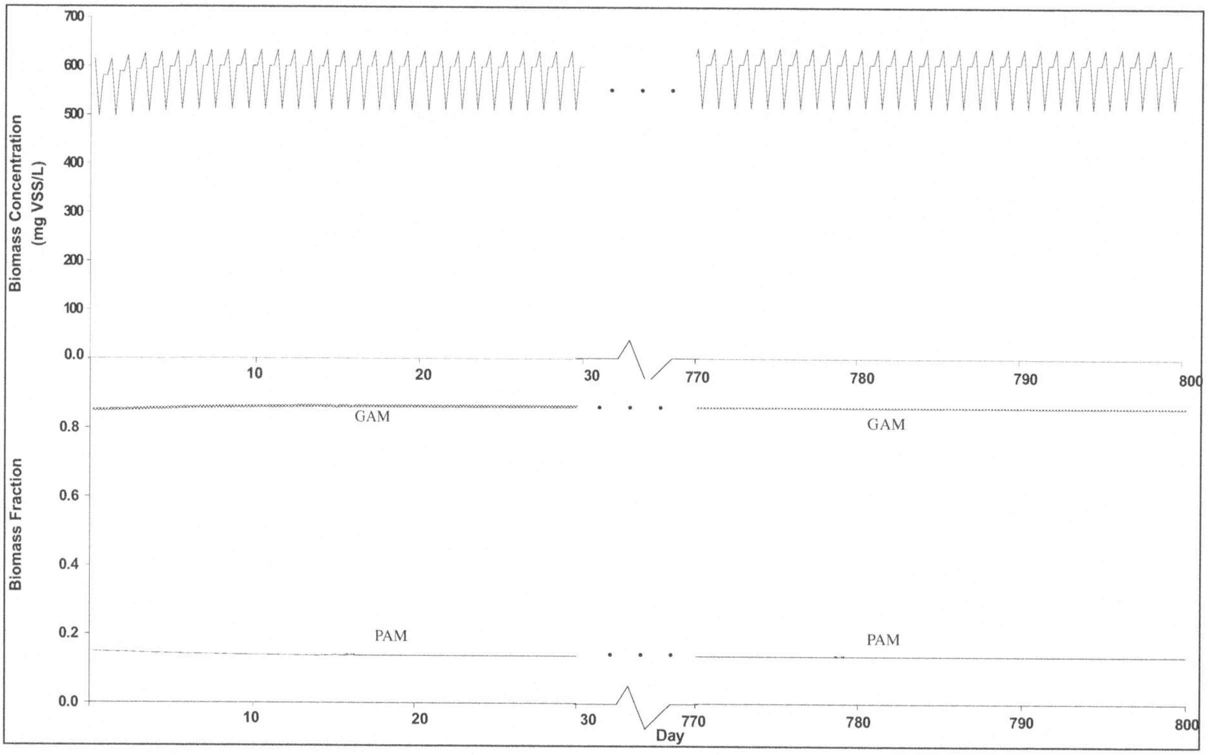

Figure 3-77. The AnA Total Biomass Concentration and PAM/GAM Biomass Fraction Profile 800-Day Simulation (Loading Pattern 1).

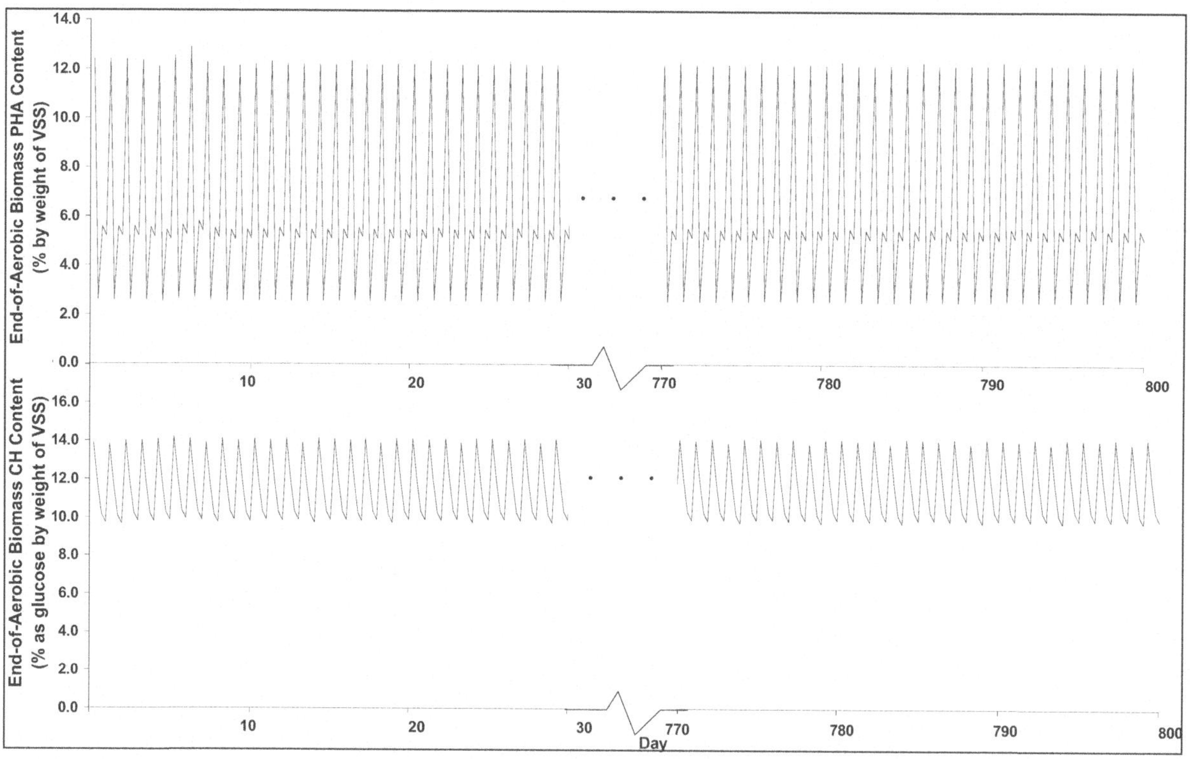

Figure 3-78. The AnA Biomass PHA and CH Content Profile, 800-Day Simulation (Loading Pattern 1).

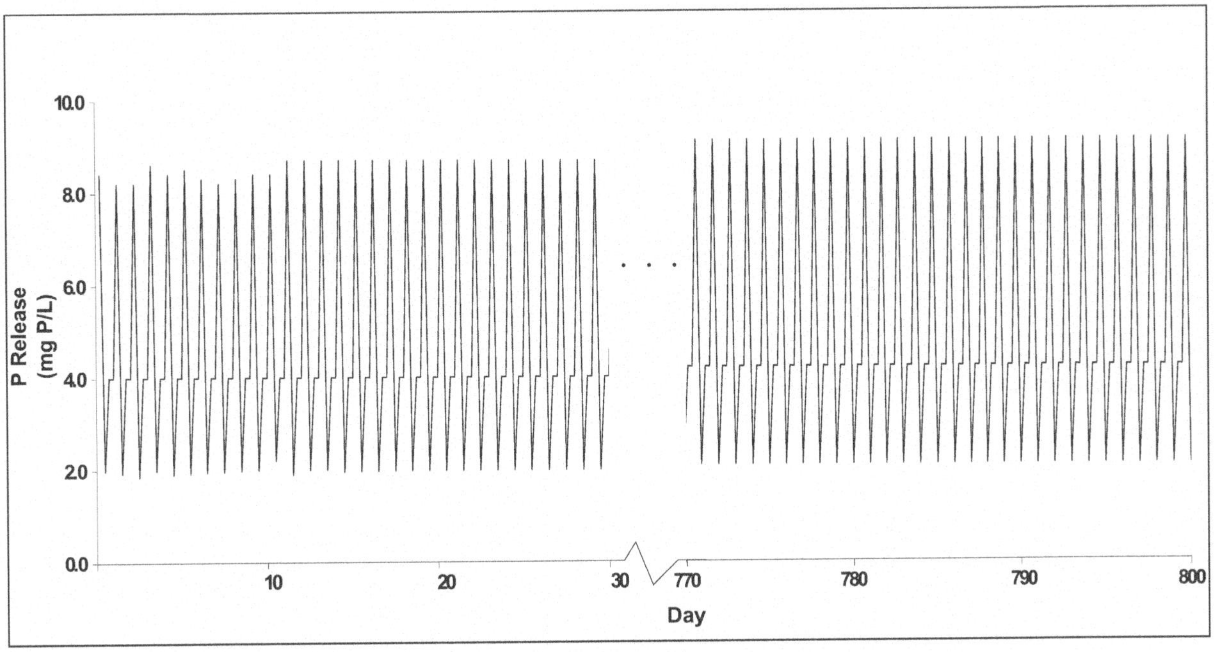

Figure 3-79. The AnA P Release Profile, 800-Day Simulation (Loading Pattern 1).

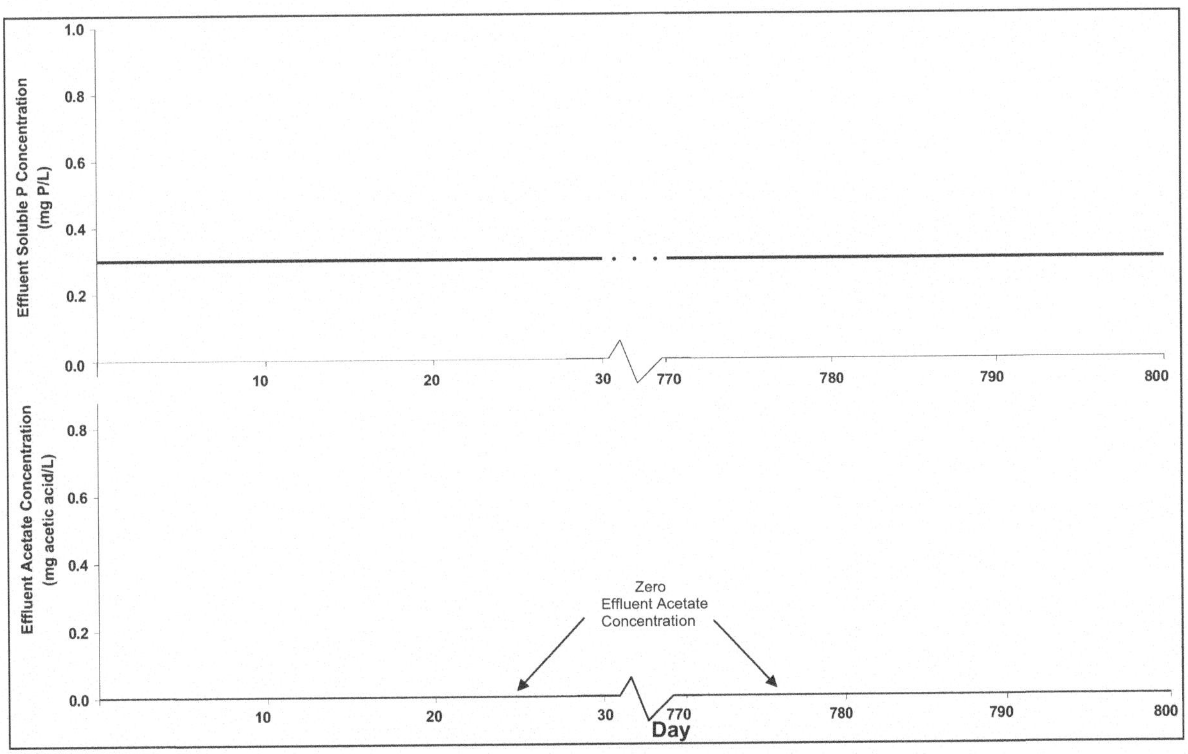

Figure 3-80. The AnA Effluent Soluble P and Acetate Concentration Profile.

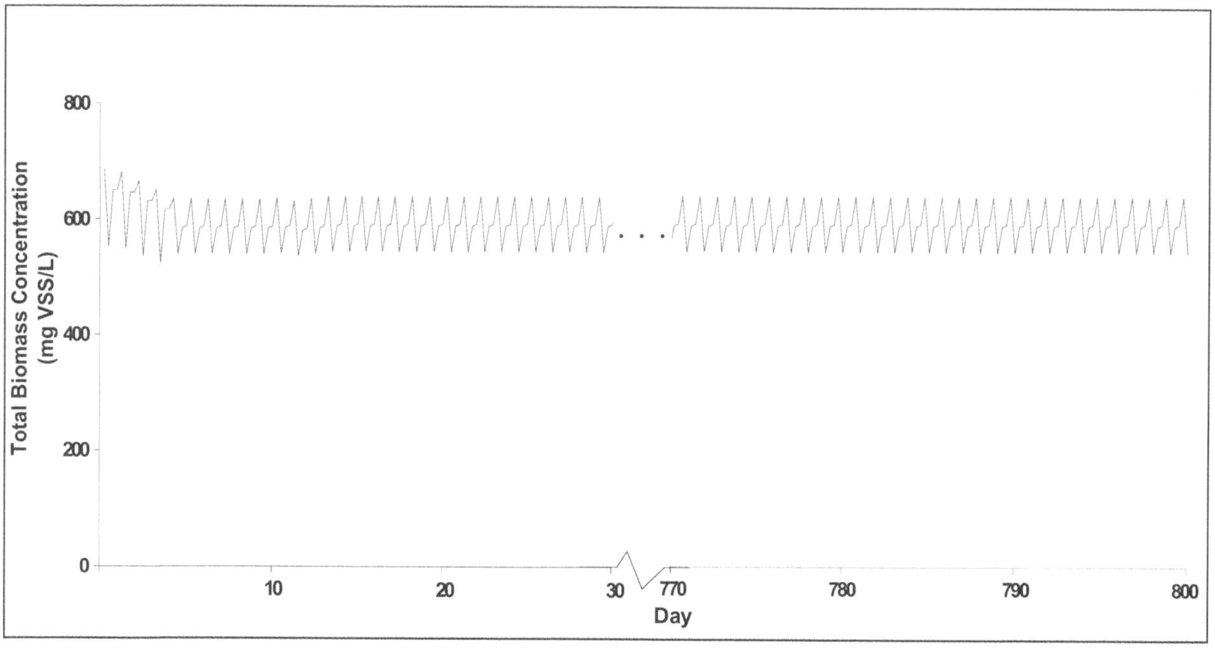

Figure 3-81. The CA Total Biomass Concentration Profile, 800-Day Simulation (Loading Pattern 1).

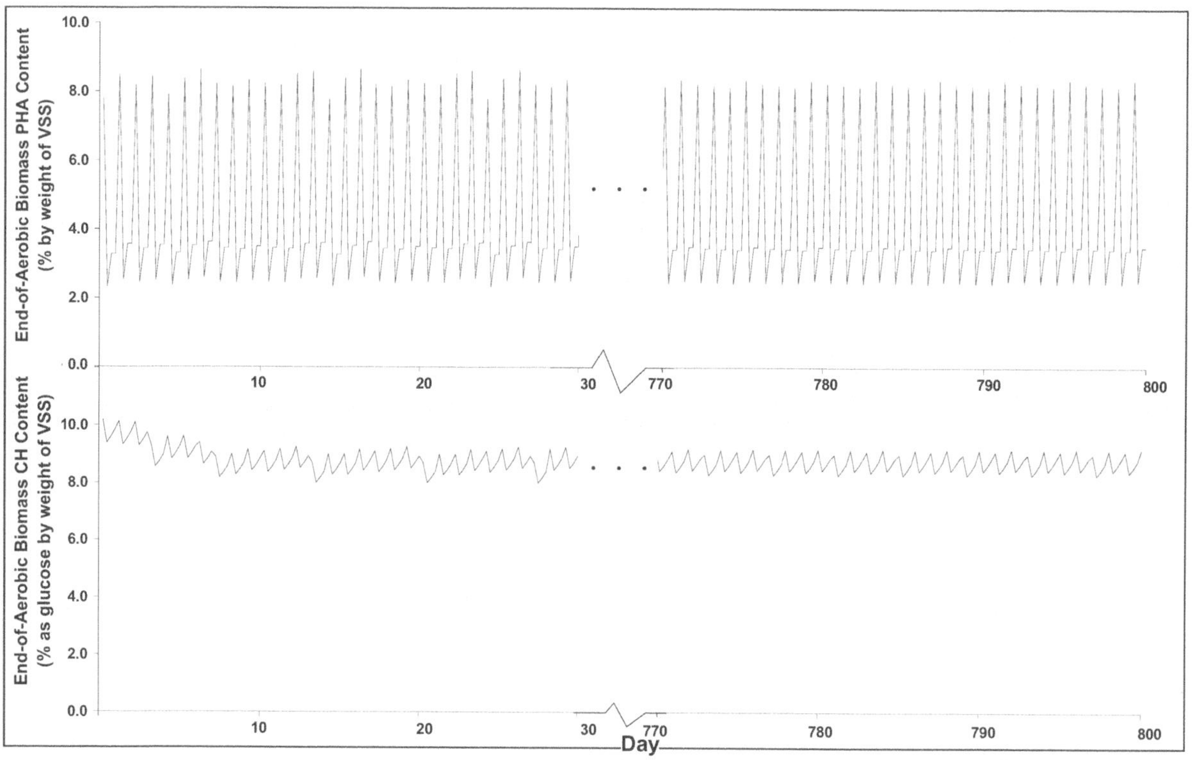

Figure 3-82. The CA Biomass PHA and CH Content Profile, 800-Day Simulation (Loading Pattern 1).

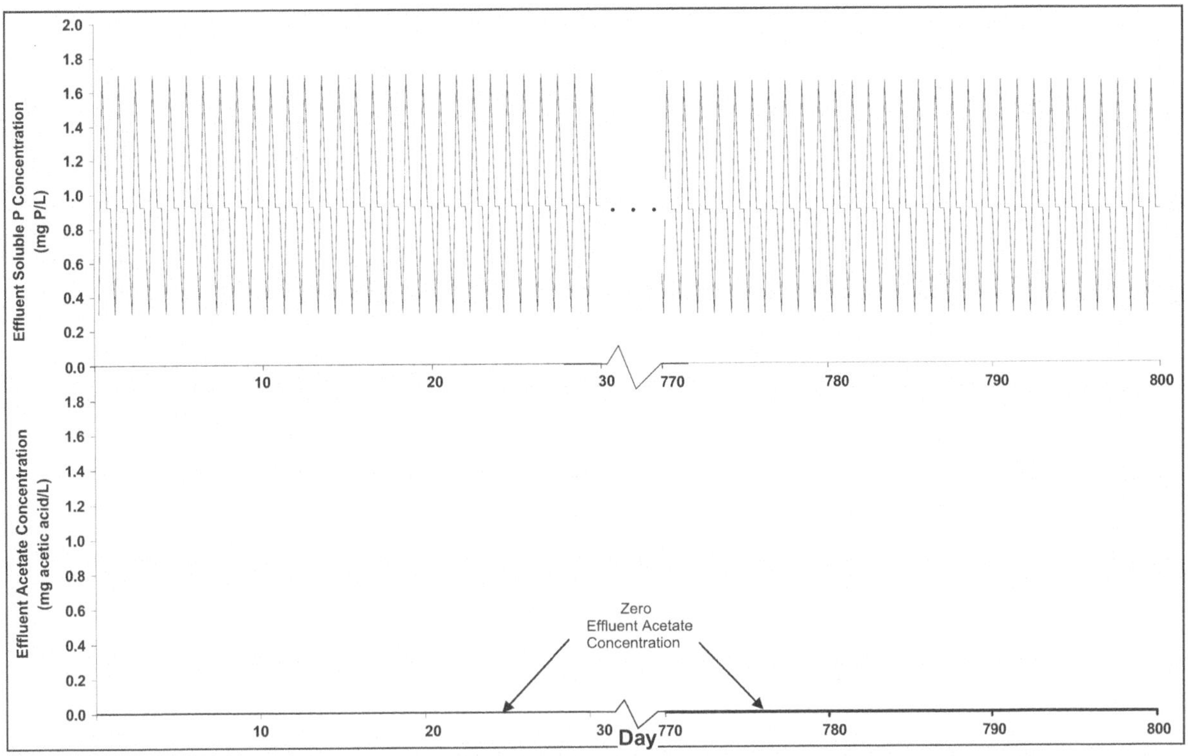

Figure 3-83. The CA Effluent Soluble P and Acetate Concentration Profile, 800-Day Simulation (Loading Pattern 1).

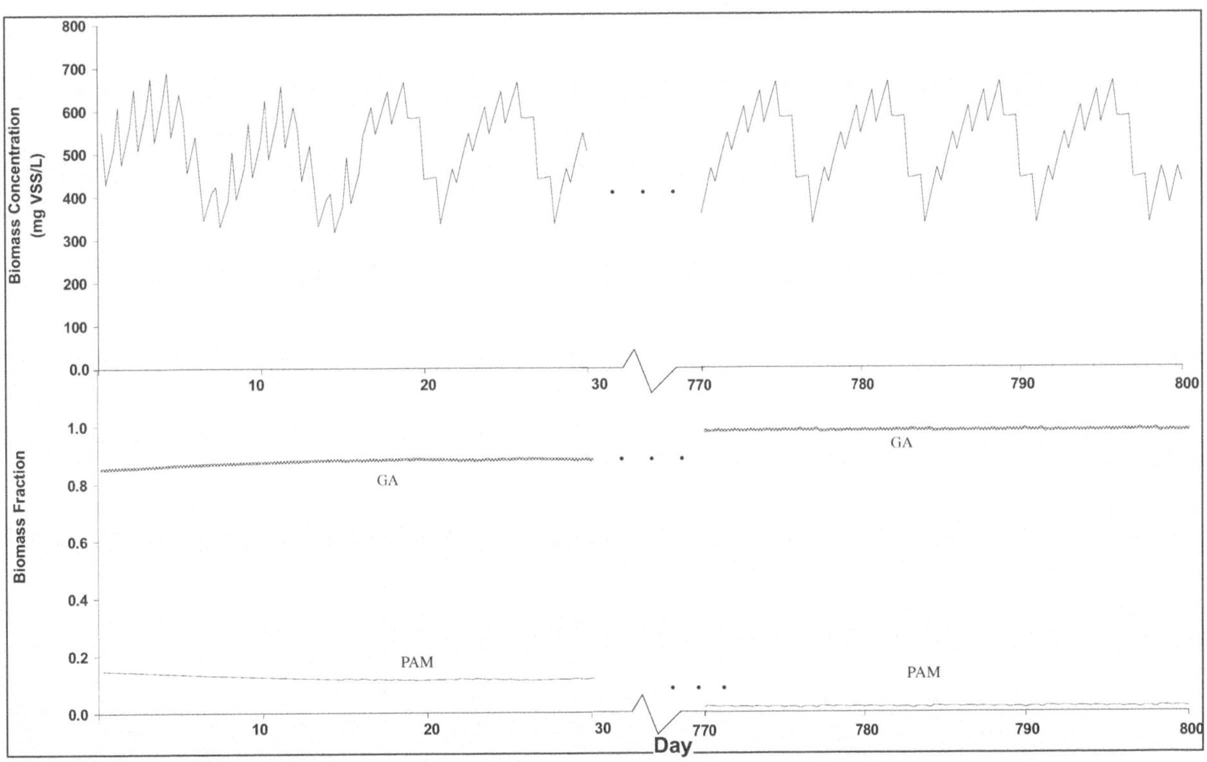

Figure 3-84. The AnA Total Biomass Concentration and PAM/GAM Biomass Fraction Profile, 800-Day Simulation (Loading Pattern 2).

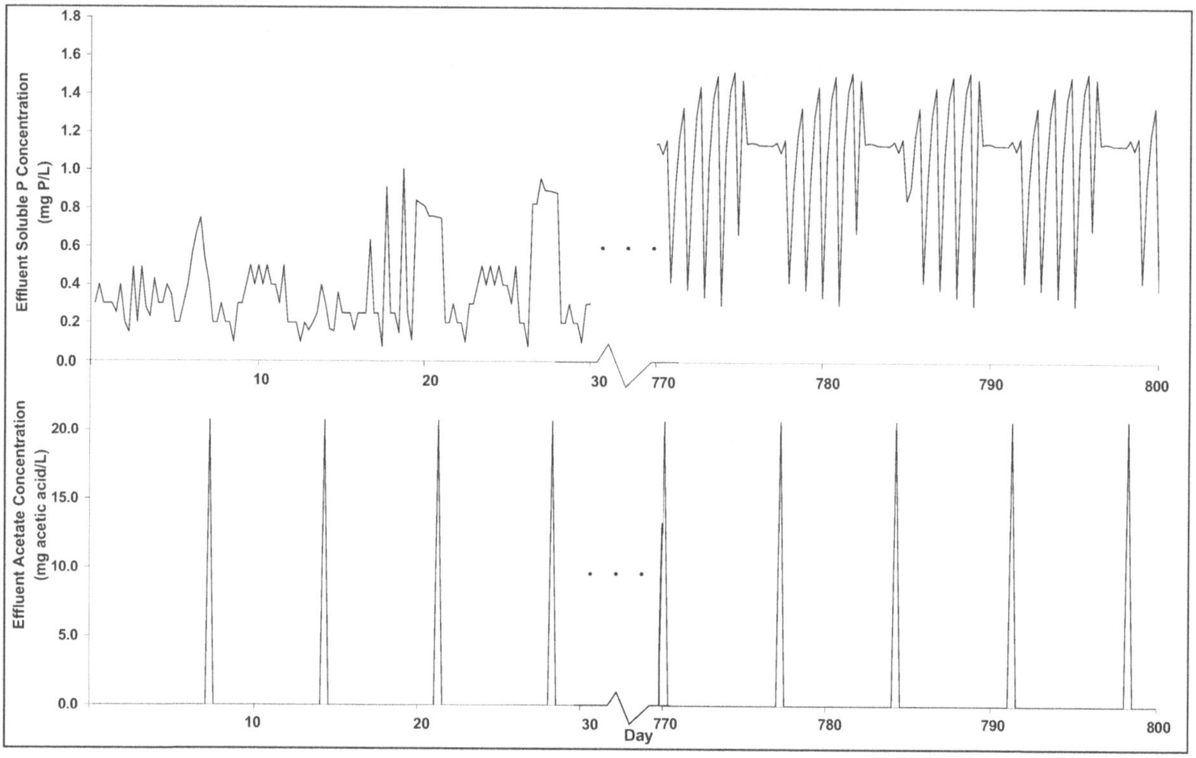

Figure 3-85. The AnA Effluent Soluble P and Acetate ConcentrationProfile, 800-Day Simulation (Loading Pattern 2).

3.5.6 Conclusions

Multiple-cycle transient COD loading experiments showed that the presence of an anaerobic zone affected the effluent quality and sludge content of AnA and CA SBRs when they were subjected to influent transient COD loading patterns that included high and low COD loading periods.

During the Loading Pattern 1 experiment, the AnA SBR maintained an average effluent soluble P concentration of ≤0.8 mg P/L while that of the CA SBR varied from approximately 0.2 mg P/L (during the high COD loading cycle) to 1.9 mgP/L (during the low COD loading cycle). The overall average soluble effluent P concentrations were 0.4 mgP/L for the AnA SBR and 1.0 mgP/L for the CA SBR. The AnA SBR anaerobically released an average of 9.2 mgP/L during the high COD loading cycle and 1.2 mgP/L during the low COD loading cycle. The AnA VSS was higher in PHA and CH contents than the CA VSS. This suggests that a greater fraction of influent COD was used for synthesis of non-P containing storage products in the AnA process during transient COD loading periods. The acetate uptake rates of both SBRs remained relatively constant. Investigators did not note any significant differences in the end-of-aerobic VSS, effluent TSS, or SVI between the systems.

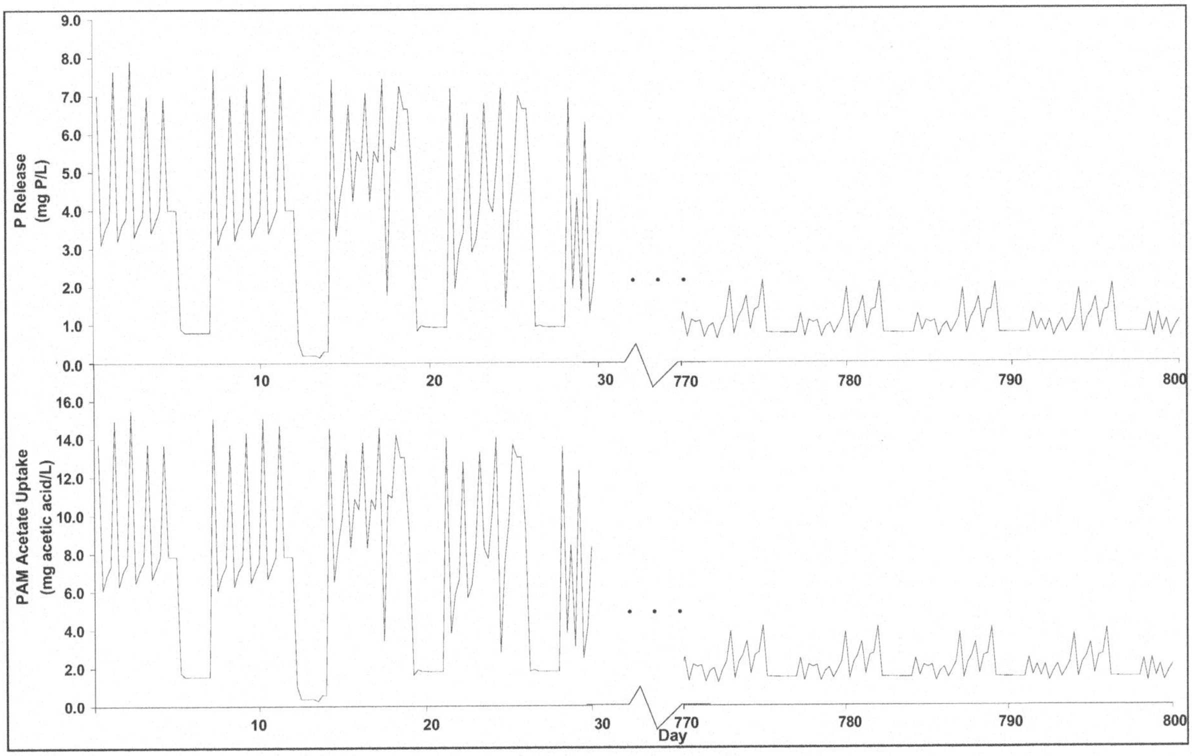

Figure 3-86. The AnA P Release and PAM Acetate Uptake Profile, 800-Day Simulation (Loading Pattern 2).

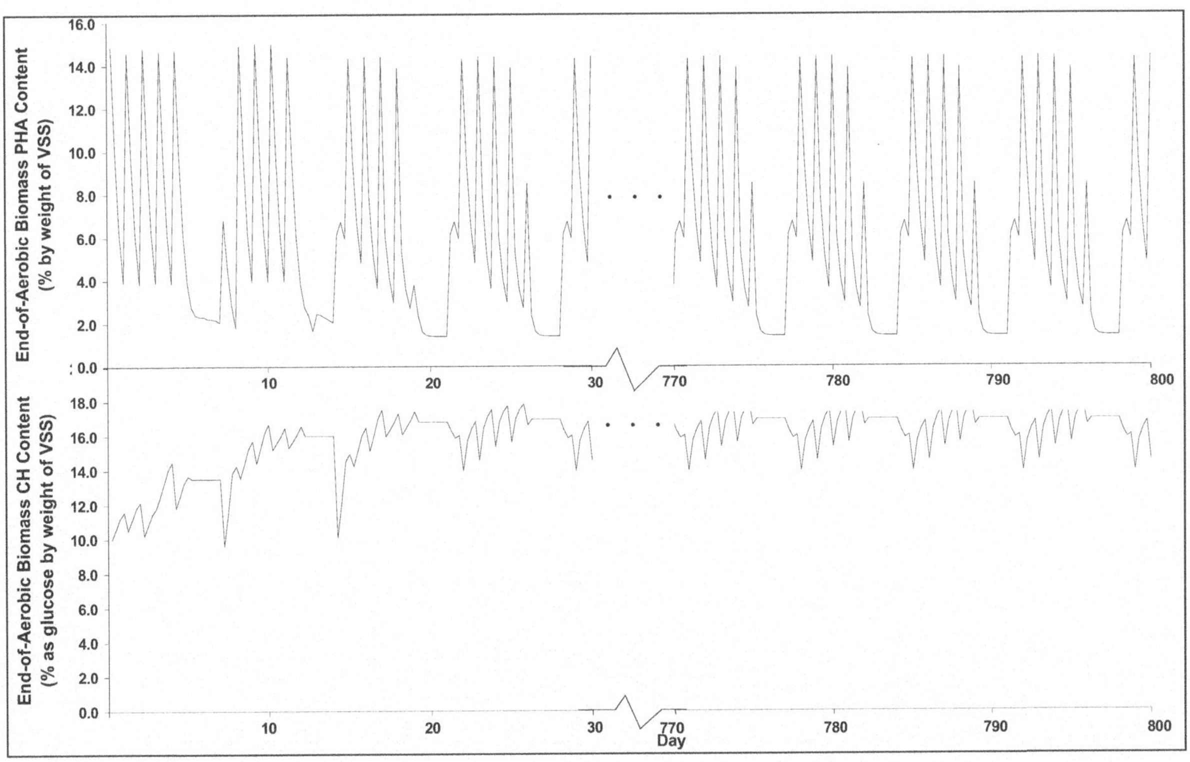

Figure 3-87. The AnA End-of-Aerobic Biomass PHA and CH Content Profile, 800-Day Simulation (Loading Pattern 2).

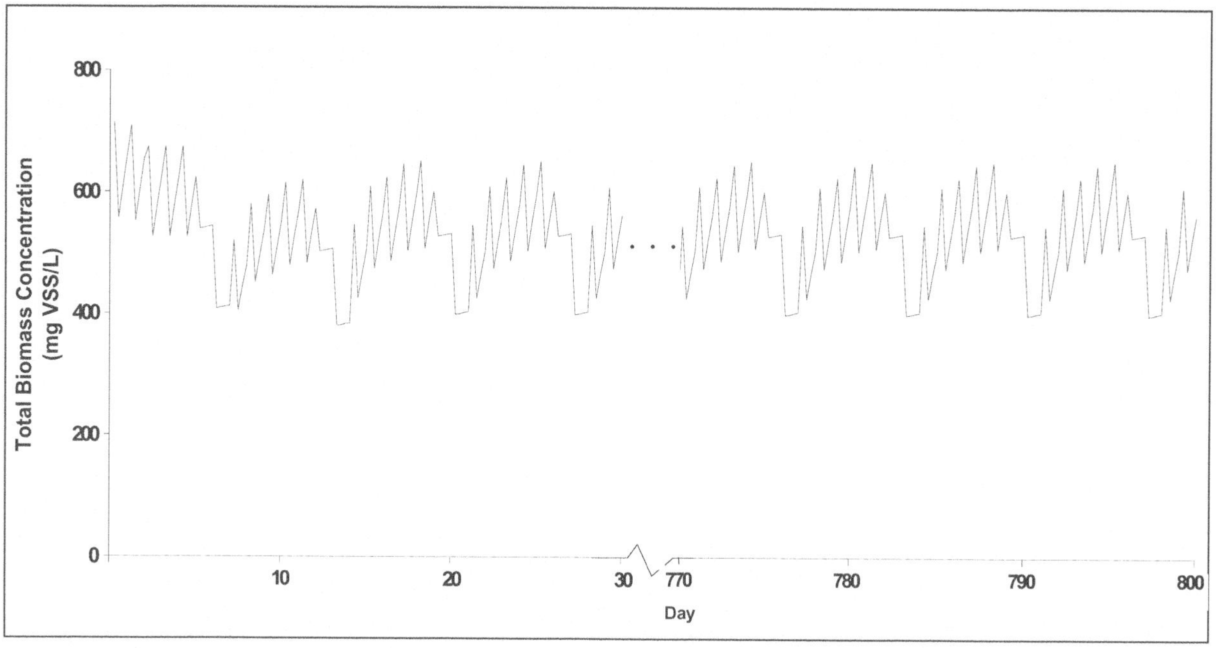

Figure 3-88. The CA Total Biomass Concentration Profile, 800-Day Simulation (Loading Pattern 2).

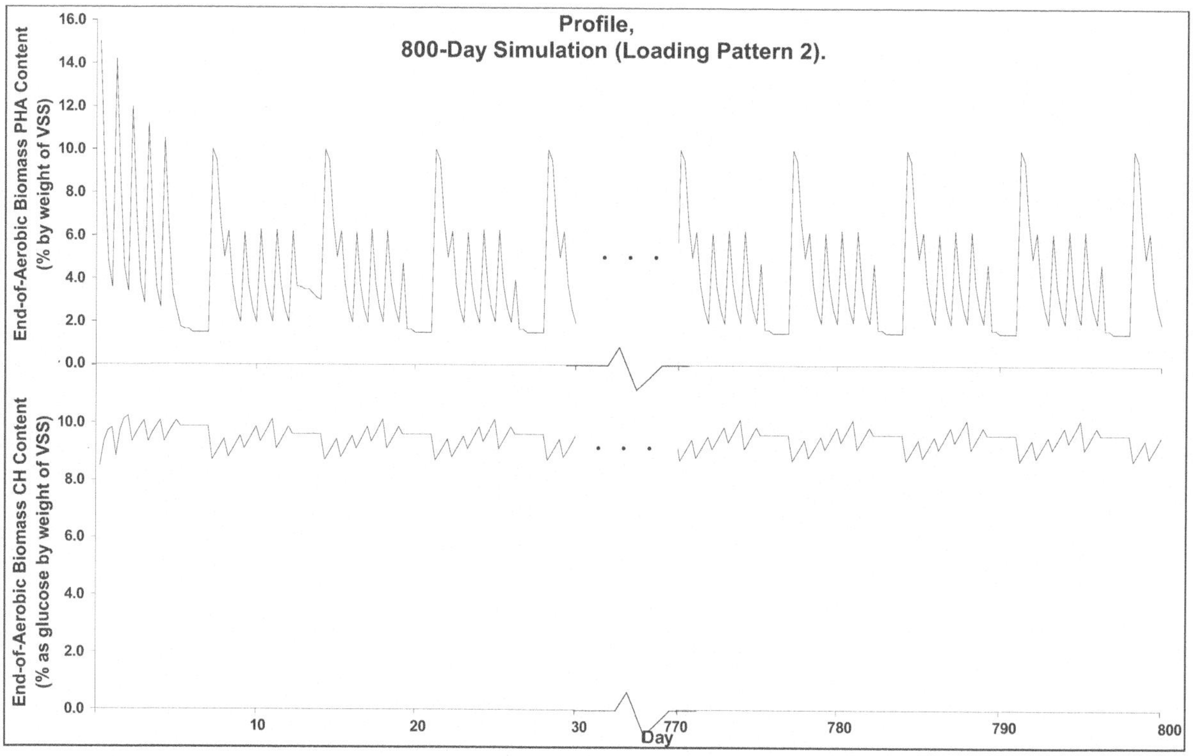

Figure 3-89. The CA End-of-Aerobic Biomass PHA and CH Content Profile, 800-Day Simulation (Loading Pattern 2).

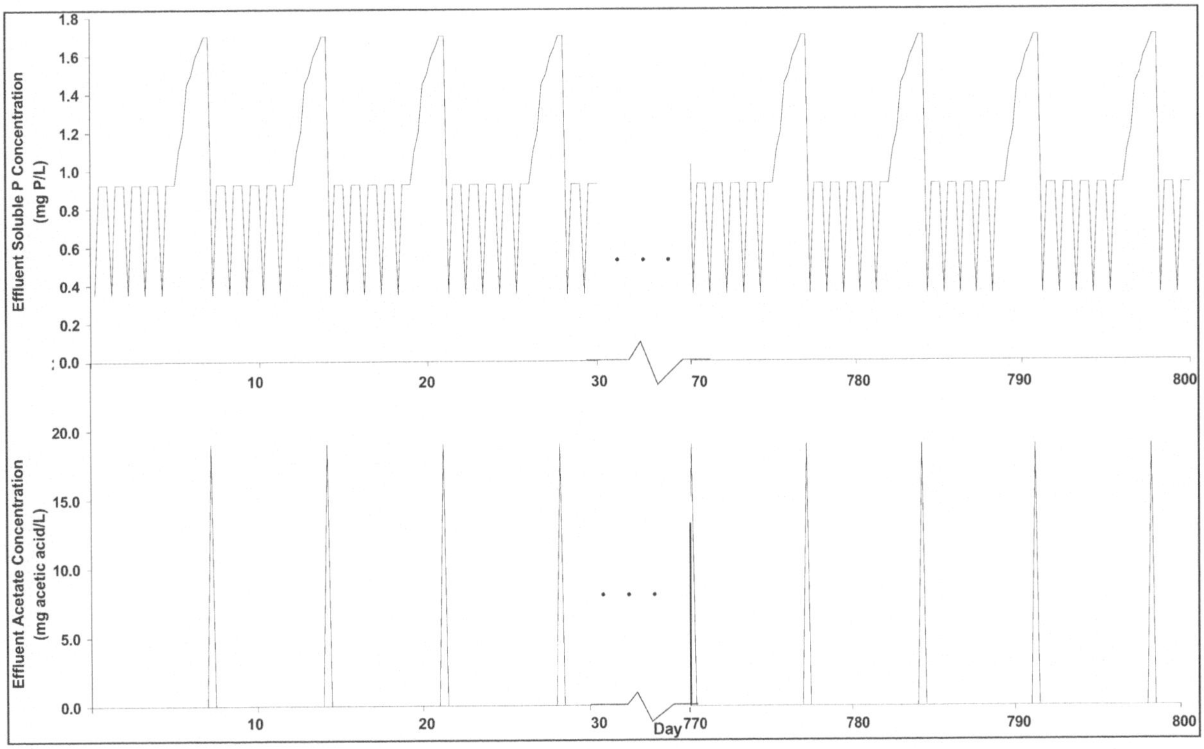

Figure 3-90. The CA Effluent Soluble P and Acetate Concentration Profile, 800-Day Simulation (Loading Pattern 2).

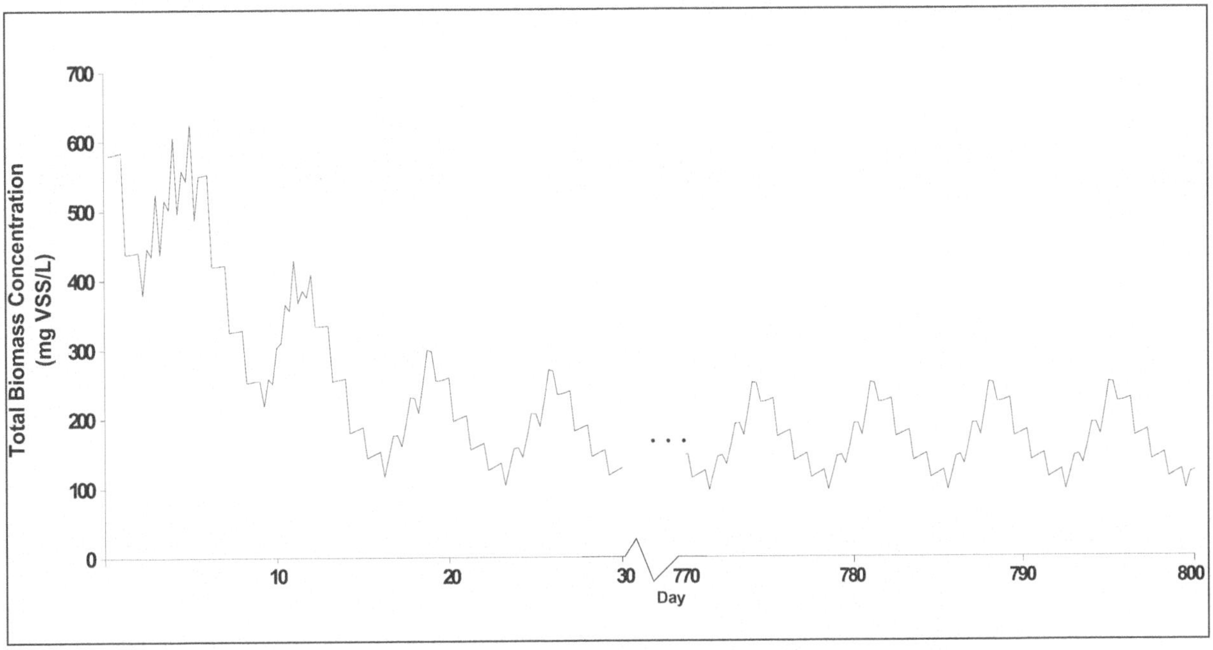

Figure 3-91. The AnA Total Biomass Concentration Profile, 800-Day Simulation (Loading Pattern 3).

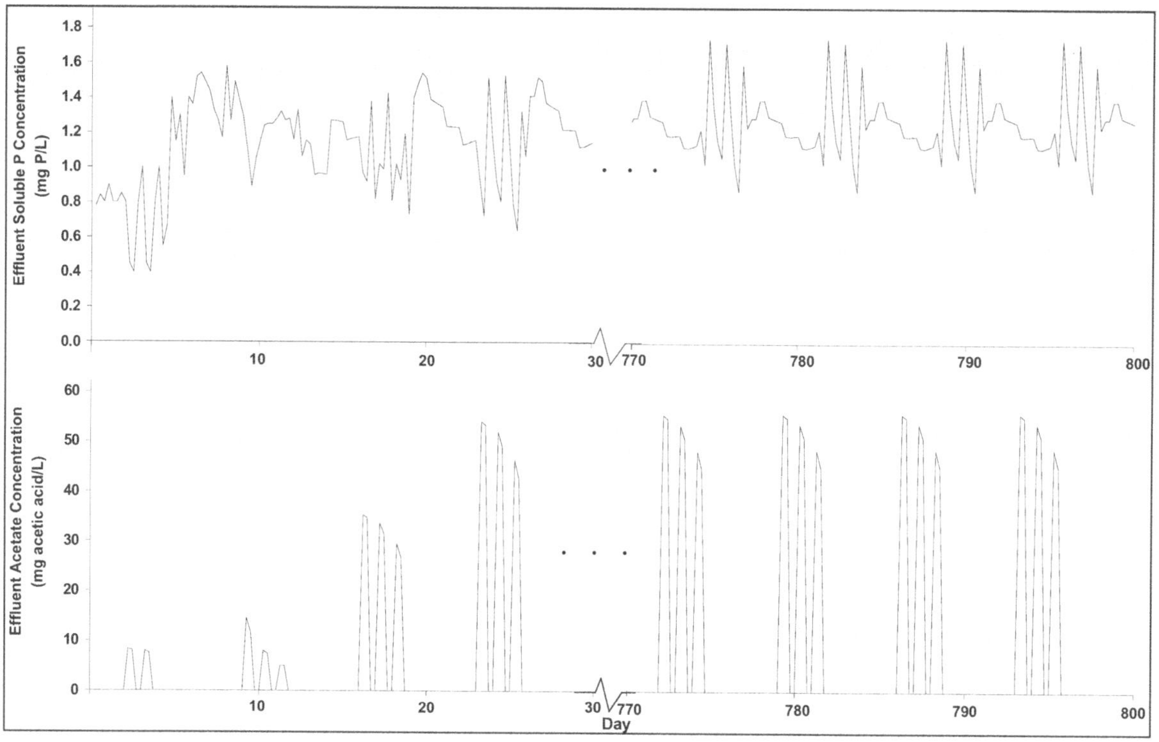

Figure 3-92. The AnA Effluent Soluble P and Acetate Concentration Profile.

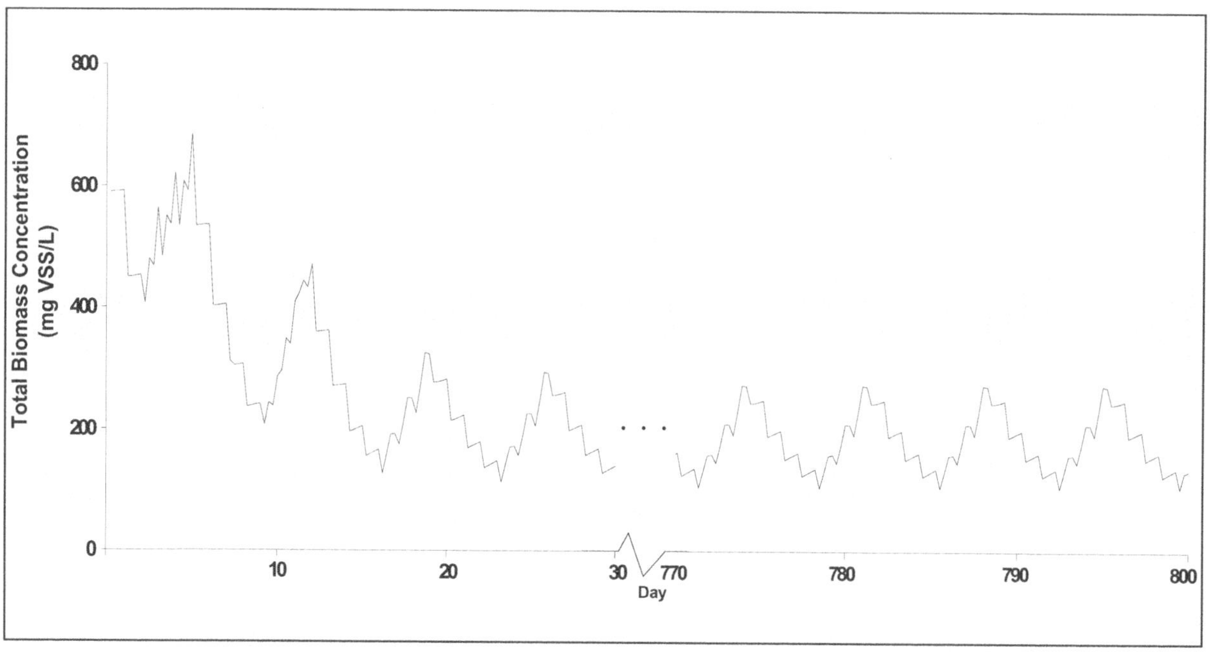

Figure 3-93. The CA Total Biomass Concentration Profile, 800-Day Simulation (Loading Pattern 3).

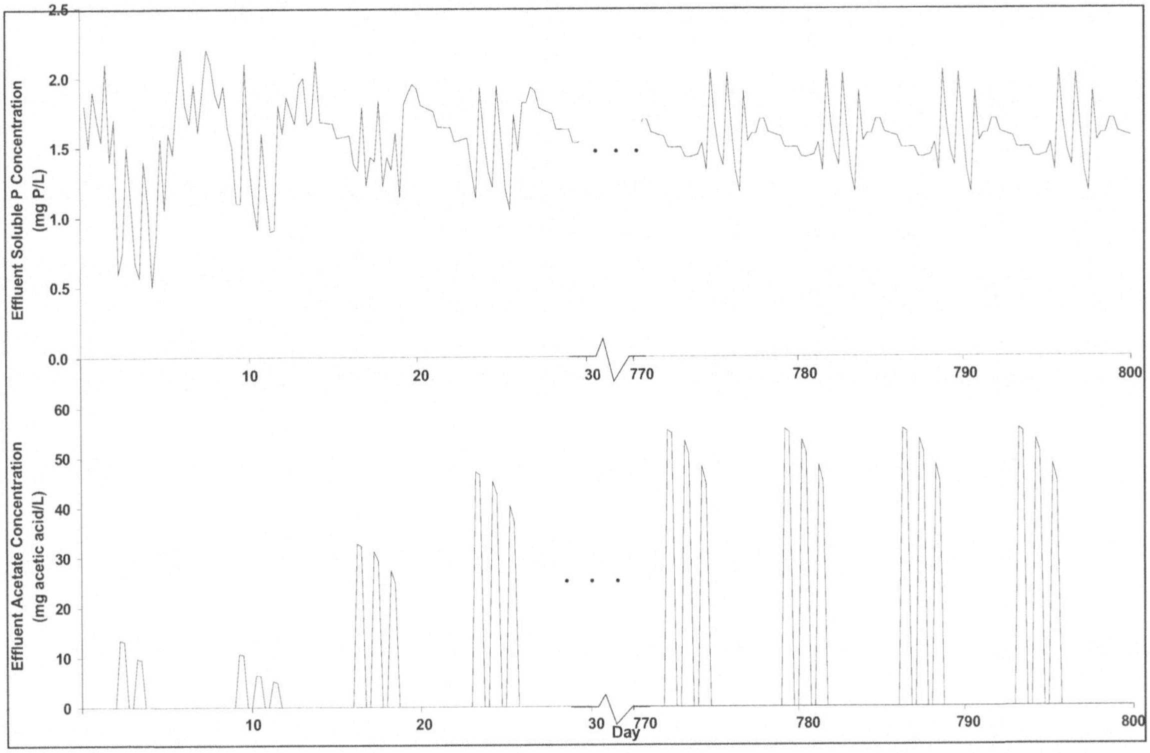

Figure 3-94. The CA Effluent Soluble P and Acetate Concentration Profile, 800-Day Simulation (Loading Pattern 3).

During the Loading Pattern 2 experiment, the CA effluent soluble P concentration averaged 0.8 mg P/L during weekday and 1.3 mgP/L during weekend COD loading periods. The AnA effluent soluble P concentration averaged 0.3 mg P/L during weekday periods, and 0.4 mgP/L during weekend COD loading periods. The weekend low COD loading period caused CA effluent soluble P concentrations to increase to 1.8 mgP/L, while the AnA SBR increased only to 0.8 mgP/L. The AnA process maintained low soluble effluent P concentration when subjected to an influent transient COD loading pattern that included a weekend low COD loading period. The AnA sludge released approximately 7 mgP/L during the high loading cycles and approximately 0.5 mgP/L during the low COD loading periods. The weekend low COD loading period resulted in a decrease in the biomass content of both SBRs, and effluent acetate was detected for both SBRs after the first high loading cycle following a low COD loading period. The AnA sludge maintained a higher PHA content than the CA sludge, while the CH content of the sludge types was similar, indicating that more non-P containing storage products were synthesized in the AnA sludge. The acetate uptake rates of both SBRs remained relatively constant during the experiment, and no discernable difference in effluent TSS and SVI was observed.

During the Loading Pattern 3 experiment, VSS contents of both SBRs dropped sharply as a result of the extended low COD loading period (4 d of low COD loading per week). The soluble effluent P concentration of both SBRs exceeded 1 mgP/L, and acetate was detected in the effluent streams of both reactors. The effluent TSS and the SVI values of both systems rose during the extended low COD loading period, indicating that the AnA and CA processes fail during an extended low COD loading period.

Long-term AnA model simulations suggested that the PAM and GAM biomass fractions were stable when treating Loading Pattern 1, but that the PAM gradually failed when treating Loading Pattern 2 because the PAM gradually captured less acetate during the first high COD loading cycle following the weekend low COD loading period. This model prediction is supported by experimental data that show a reduction in P release during the first high COD loading cycle following the weekend low COD loading period (Figure 3-66). Simulations also showed that the AnA sludge failed when subjected to Loading Pattern 3 due to the reduction in biomass caused by the 4-d low COD loading period. CA simulations showed stable VSS and storage product levels when treating Loading Patterns 1 and 2, but showed elevated P and acetate concentrations when treating Loading Pattern 3 due to the reduction in biomass caused by the 4-d low COD loading period.

CHAPTER 4.0

Practical Significance of Research Results

The following practical significant conclusions can be drawn from the results of the research.

1. When treating wastewater containing readily biodegradable soluble organics an activated sludge system that had a high degree of plug flow produced sludge that settled far better than that from a completely mixed aeration basin with anaerobic selectors.

   This point was illustrated by the inability to produce a settleable activated sludge when treating an acetate-casamino acids based synthetic wastewater using a continuous flow system with anaerobic selectors. An activated sludge with acceptable settling properties was produced in dump-fed sequencing batch reactors (SBRs) both when the initial "react" conditions were mixed and aerated ("aerobic", CA) and when they were mixed but not aerated ("anaerobic" AnA).

2. When treating a phosphorus (P) deficient wastewater the P requirements of an actvated sludge developed in an AnA SBR were approximately 20% lower than for an activated sludge from a CA SBR.

   The research was conducted entirely at a mean cell residence time (MCRT) of 4d (largely to avoid nitrification, that would have complicated interpretation of the results). Because the P requirements of activated sludge vary with MCRT, the results of this research apply directly only to systems with a 4d MCRT. In general because net total sludge production and storage product production decrease as MCRT increases, the differences between an AnA and a CA system should also be smaller at sludge ages higher than 4d.

3. An AnA SBR activated sludge treating a P deficient wastewater was better able to withstand organic load variations (especially decreases in organic load) than a CA SBR activated sludge treating the same wastewater.

   This finding has practical significance for activated sludge systems treating P-deficient wastewaters that also have to meet stringent effluent P requirements. An AnA activated sludge has the ability to take up (and store as polyphosphate) the soluble P not required for growth. This produces an effluent with low P content during times when organic loads are low. When the P demand of an organic load exceeds the P supply the stored polyphosphate becomes available to meet this demand and prevent P deficiency. Thus the "buffering effect" of the initial anaerobic zone provides a means to produce effluents low in organic matter and P when treating a P-deficient wastewaters with variable influent loads.

APPENDIX A

CALCULATION OF CARBON DIOXIDE PRODUCTION

Carbon dioxide production during the anaerobic phase of the AnA SBR occurs as $C_{acetate}$ and CCH are converted to C_{PHB}, C_{PHV}, C_{co2} according to the following (Arun, 1988; Schuler, 1998):

$C_{acetate} + {}^{1}/_{2}$ ATP Æ $C_{acetylCoA}$ (rA-1)
$C_{CH} \rightarrow {}^{2}/_{3}\ C_{acetylCoA} + {}^{1}/_{3}\ CCO2 + {}^{2}/_{3}\ NADH + {}^{1}/_{2}\ ATP$ (rA-2)
$C_{CH} + {}^{1}/_{3}NADH \rightarrow C_{propionylCoA} + {}^{1}/_{3}\ ATP$ (rA-3)

Acetyl-CoA and propionyl-CoA are used in PHA synthesis:

$C_{acetylCoA} + {}^{1}/_{4}\ NADH \rightarrow C_{PHB}$ (rA-4)
${}^{2}/_{5}\ C_{acetylCoA} + {}^{3}/_{5}\ C_{propionylCoA} + {}^{1}/_{5}\ NADH \rightarrow C_{PHV}$ (rA-5)

The total CH degradation rate is sum of the reactions rA-2 and rA-3:

$$rC_{CH} = rA\text{-}2 + rA\text{-}3 \quad \text{(A-1)}$$

The propionyl-CoA synthesis rate can be related to the PHV synthesis rate as follows:

$$r_{CpropionylCoA} = rA\text{-}3 = {}^{3}/_{5}\ rA\text{-}5$$

Carbon dioxide production is then calculated as follows:

$$\begin{aligned} rC_{CO2} &= {}^{1}/_{3}\ rA\text{-}2 \\ &= {}^{1}/_{3}\ (rC_{CH} - rA\text{-}3) \\ &= {}^{1}/_{3}\ (rC_{CH} - {}^{3}/_{5}rA\text{-}5) \\ &= {}^{1}/_{3}\ (rC_{CH} - {}^{3}/_{5}r\ C_{PHV}) \end{aligned} \quad \text{(A-2)}$$

Equation A-2 was used to calculate anaerobic phase carbon dioxide production based on carbohydrate degradation and PHV synthesis.

Aerobic phase carbon dioxide production in the anaerobic/aerobic reactor was calculated by considering four simultaneous processes, biomass synthesis (rA-6), polyphosphate synthesis (rA-7), glycogen formation (rA-8), and cell maintenance (rA-9) as follows (Smolders, 1995):

$1.37\ C_{PHA} \rightarrow CX + 0.37\ C_{CO2}$ (rA-6)
$0.27\ C_{PHA} \rightarrow 0.27\ C_{CO2}$ (rA-7)
$1.12\ C_{PHA} \rightarrow C_{CH} + 0.12\ C_{CO2}$ (rA-8)
$C_{PHA} \rightarrow C_{CO2}$ (rA-9)

In the above expression, both PHB and PHV are assumed to participate in each reaction. The specific rate constants associated with the above reactions are kA-6 (0.15 mol/mol/h), kA-7 (0.55 mol/mol/h), kA-7 (0.5 mol/mol/h), and kA-8 (0.004 mol/mol/h). Thus the total C_{CO2} production was calculated using the stoichiometric ratios and kinetics constants as follows:

$$rC_{CO2} = rC_{PHA} * [(0.37/1.37)*(0.15/1.5) + (1)*(0.55/1.5) + (0.12/1.12)*(0.5/1.5) + (1)*(0.004/1.5)] \quad \text{(A-3)}$$

Carbon dioxide production from the completely aerobic SBR during the first 110 minutes was calculated by considering CH degradation and PHA synthesis (as shown previously for the anaerobic phase of the AnA SBR), and biomass growth. The carbon dioxide production rate due to CH degradation and PHA synthesis is as shown in equation A-2 (eqA-2), and the carbon dioxide production rate due to biomass growth is as shown in reaction A-6 (rA-6). Thus the total carbon dioxide production rate was calculated as follows:

$$rC_{CO2} = {}^{1}/_{3}\,(rC_{CH} - {}^{3}/_{5}\,rC_{PHV}) + rC_{X}*\,(0.37) \quad \text{(A-4)}$$

Carbon dioxide production after the first 110 minutes of the cycle was calculated based on biomass growth, carbohydrate synthesis, and maintanence according to the following:

$$rC_{CO2} = rC_{X}*\,(0.37) + 0.12*rC_{CH} + 0.004*C_{X} \quad \text{(A-5)}$$

APPENDIX B

AnA and CA Model Development, Parameter Values, and Program Code (Matlab 5.3 Platform)

Models for the AnA and CA SBRs were used to simulate growth of cells, removal of acetate and P, and transformation of biomass storage products. Modeling was conducted using a Matlab 5.3 (The Mathworks, Inc., Natick, MA) platform. The steps used in model development are described below. Note that the development of the AnA model included assessment of the coexistence of two competing metabolisms (PAM and GAM), whereas the CA model simulated overall sludge activity.

B.1 AnA Model Development

B.1.1 Development of Mathematical Expressions

The expressions used were similar to those in the Smolders Model (Smolders et. al, 1995), which included a description of PAM. The Smolders model was used because it was developed using fundamental biochemical pathways and should therefore be more widely applicable than empirically-based models. The model rate expressions are shown in Table B-1. The kinetic expressions used for PAM were initially assigned as in the Smolders model. The kinetic expressions for the GAM were developed from the literature. GAM includes the same anaerobic and aerobic processes in PAM, except those that involve PP synthesis and degradation. Initial values for model parameters were selected from the literature.

Table B-1. AnA Model Rate Expressions.

Process	Specific Rate Expression (mole/mole/h) and variables (mM)	Parameters PAM(p), GAM(g)	Note
Anaerobic Substrate Uptake	$q_{smax} * [C_s/(C_s+K_s)]$ C_s = mM acetate-C	pq_{smax}, gq_{smax} pK_s, gK_s	1
Aerobic Biomass Synthesis	$k_x (C_{pha}/C_x)$ C_{pha} = mM PHA-C C_x = mM biomass-C	[3]pk_x, [3]gk_x	2
Aerobic Phosphate Uptake	$k_{pp}*[C_p/(C_p+K_p)]*(1-C_{pp}/C_{ppmax})*C_{pha}/C_x$ C_p= mM P C_{pp}= mM P	k_{pp}, k_p $C_{ppmax} = 0.4*C_{PAM}$	2
Aerobic Glycogen Formation	$k_{gl}*[(C_{glymax}-C_{gly})/C_x]$ C_{gly} = mM GLY-C C_x = mM biomass-C	pk_{gl}, gk_{gl} $pC_{glymax} = 0.34*C_{PAM}$ $gC_{glymax} = 0.34*C_{GAM}$	2
Maintenance	m	p_m, g_m	2

[1] Stoichiometrically linked to anaerobic phase PHA, CH, and PP rate expressions
[2] Stoichiometrically linked to aerobic phase PHA rate expression
[3] Target Parameter

B.1.2 Preliminary Model Evaluation

A preliminary model was evaluated to determine which parameters had most influence on whether coexistence of the PAM and GAM was predicted by the model. A series of model simulations were performed in the model parameters shown in Table B-1 were varied (pqs_{max} and gqs_{max} were not included in this analysis due to availability of literature values and experimental data). The parameters that had the strongest influence on PAM and GAM coexistence were used in the sensitivity analysis. These target parameters were the maximum specific growth rates for the PAM (pk_x) and GAM (gk_x) (also indicated in Table B-1).

B.1.3 Sensitivity Analysis

In the sensitivity analysis, the target model parameters were systematically varied, and relative concentrations over time of PAM and GAM biomass were determined to see whether coexistence of the two metabolisms was achieved. Using this approach, a profile was developed that shows which combinations of target parameters yielded stable behavior (coexistence). This combination of target parameters is referred to as the "stable region." The stability profile is shown in Figure B.1.

B.1.4 Calibration

The purpose of model calibration was to determine which target parameter values should be used for model simulation. The stable region shown in Figure B-1 was inspected. Simulations were performed using target parameter values in the stable region, and the values producing results that agree most closely with experimental data were identified. Table B-2 shows the results of the model calibration. The target parameter values producing the lowest total root mean square error resulted in the model simulation that most closely agreed with experimental data. The parameter values selected for process simulation were pk_x = 0.2 M biomass-C/M PHA-C/h, gk_x = 0.2 M biomass-C/M PHA-C/h. AnA model parameter values and stoichiometry are shown in Tables B-3 and B-4.

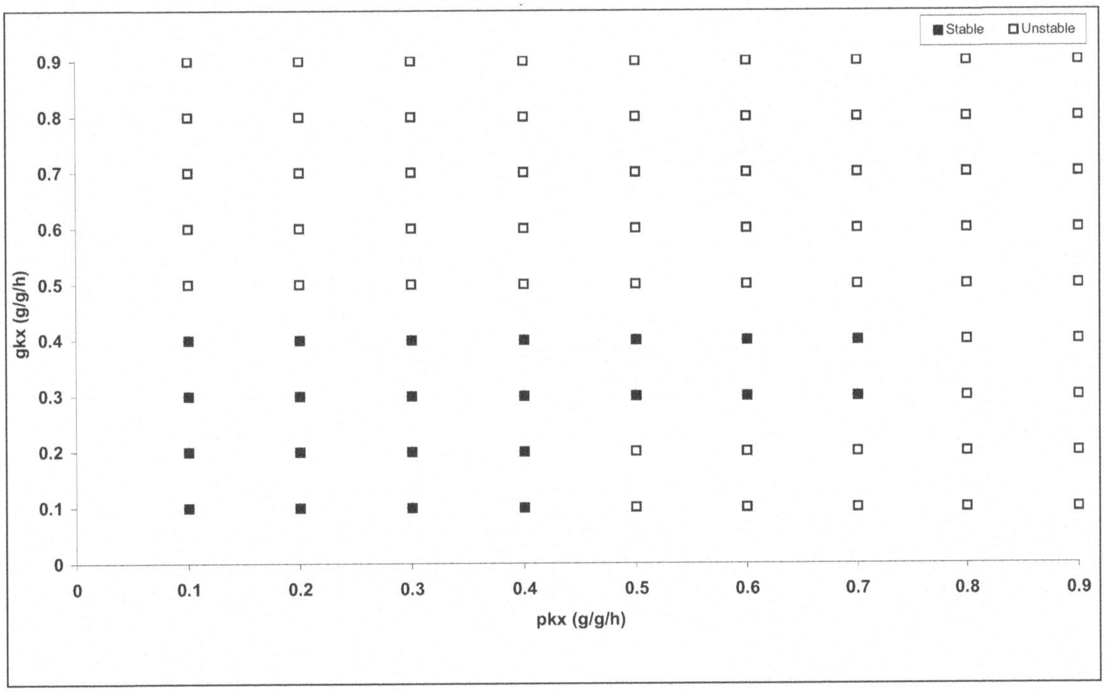

Figure B-1. Stability Profile – AnA Model.

Table B-2. AnA Model Calibration Results.
(Five parameter combinations with lowest total root mean square error shown)

Parameter Combination (pk_x,gk_x)	Error in Anaerobic P Release Conc., %	Error in Total EOA VSS Conc., %	Error in EOA PHA Conc., %	Error in EOA CH Conc., %	Error in Acetate Uptake Conc., %	Total Root Mean Square Error[1], %
(0.2,0.2)	4.7	-3.3	-25	1	1.3	**26**
(0.3,0.2)	23	-13	-63	5.5	2.3	69
(0.1,0.2)	10.5	1.6	-78	6.5	0.8	79
(0.2,0.3)	13.8	1.7	-78	6.5	0.8	79
(0.1,0.3)	15.8	0.2	-100	0.4	2.3	101

[1]Total root mean square error calculated as the square root of the sum of the squares of the errors in the 5 parameters shown (i.e. $[(a)^2 + (b)^2 + (c)^2 + (d)^2 + (e)^2]^{(0.5)}$)

Table B-3. AnA Model Parameter Values.

Parameter	Value	Unit	Reference
pk_x	0.2	mol C/mol C/h	1
gk_x	0.2	mol C/mol C/h	1
pq_{smax}	0.26	mol C/mol C/h	1
gq_{smax}	0.16	mol C/mol C/h	1
pK_s	1.6	mol C/L	2
gK_s	1.6	mol C/L	2
K_{pp}	0.55	mol P/L	2
k_p	0.1	mol P/mol C/h	2
p_{kgl}	0.6	mol C/mol C/h	1
g_{kgl}	0.5	mol C/mol C/h	1
p_m	0.004	mol C/mol C/h	3
g_m	0.004	mol C/mol C/h	3

[1]Current Research
[2]Smolders et. al. (1995)
[3]Filipe and Diagger (1998)

Table B-4. AnA Model Stoichiometry.

Ratio	Coefficient	Reference
P Release/Acetate-C Removed	variable[1]	Current research
PHA-C Synthesized/Acetate-C Removed	1.2 (PAM) variable[1] (GAM)	Current research
CH-C Degraded/Acetate-C Removed	variable[1]	Current research
PHA-C Degraded/ Polyphosphate-P Synthesized	0.6	Current research
PHA-C Degraded/ GLY-C Synthesized	1.3	2

[1]A function of Influent COD Peak Load Factor (See section 3.4.3.).
[2]Smolders et. al. (1995)

B.2 AnA Model Program Code

B.2.1 Specification of Parameters

```
%This program calls the anaerobic/aerobic SBR program (go6) after specification of parameters
tstart = clock;

%global declarations
global pqsmax gqsmax default initconditions count fc pfraction pkx gkx ze1 ze2 gks pks ICPLF
ico
global checktot1 checkpam1 checkgam1 checktot checkpam checkgam
global aaa bbb ccc ddd eee fff ggg hhh iii jjj kkk lll pamy gamy IntP10 pfglmax gfglmax pkgl
gkgl

%value assignments
tolerance = 0.5;
default = 1;
initconditions = 2;
countt = 1;
pfraction = 0.15;
gkx = 0.20;
pqsmax = 0.26;
gqsmax = 0.15;
pkx = 0.20;
gks = 1.6;
pks = 1.6;
pamy = 0.28;
gamy = 0.28;
IntP10 = 1.1;
pfglmax = 0.2;
gfglmax = 0.2;
pkgl = 0.55;
gkgl = 0.45;
counter = 9;
stable = 0;

go6

tstop = clock;
runtime = etime(tstop,tstart);
runtime = runtime/60;
disp(runtime)
disp('minutes')
%end
```

B.2.2 Reactor Operation and Graphing

```
%The BIOP30 M-file has both the PAM and GAM included

%The parameters in the matrix are respectively:
%C-HAc (1) P-PO4 (2) Biomass (3) C-PHA (4) PP (5) C-Gly (6)
%C-PAMbiomass (7) C-GAMbiomass (8) C-PAMpha (9) C-GAMpha (10) C-PAMgly (11) C-
GAMgly (12)

%global declarations
global pqsmax gqsmax default initconditions count fc pfraction pkx gkx gks pks i ICPLF ico
global checktot1 checkpam1 checkgam1 checktot checkpam checkgam pamy gamy IntP10
pfglmax gfglmax pkgl gkgl

%process parameters
hh = 0;
anHRT = 1.8;
aerHRT = 3;
MCRT = 4;
jj = 1/MCRT;
j = 1 - jj;
IntHAc10 = 94;
IntHAc = IntHAc10/30;
IntP = IntP10/31;
Trans = 1;
cycles = 140;
brk = 84;

%Time intervals
to = 0;
tf = anHRT+aerHRT;
t1 = anHRT;
t2 = to + .4;
t3 = t2 + .4;
t4 = t3 + .4;
t5 = t4 + .4;
t6 = t5 + .4;
t7 = t6 + .4;
t8 = t7 + .4;
t9 = t8 + .4;
t10 = t9 + .4;
TSPAN = [to tf];

%Initial Conditions
intbiomass = 20;
%Initial intbiomass = 20;
```

```
inttotpha = 1.0;
intpp = 0.25;
inttotgly = 4.0;

intpam = intbiomass*pfraction;
intgam = intbiomass - intpam;
intpampha = inttotpha*pfraction;
intgampha = inttotpha - intpampha;
intpamgly = inttotgly*pfraction;
intgamgly = inttotgly - intpamgly;

%high/low combination transient
%LFD = [0.2 1.0 1.0 1.8 0.2 1.0 1.0 1.8 0.2 1.0 1.0 1.8 0.2 1.0 1.0 1.8 0.2 1.0 1.0 1.8 0.2 1.0
1.0 1.8 0.2 1.0 1.0 1.8];

%weekend low loading transient
%LFD = [1.0 1.0 1.0 2.4 1.0 1.0 1.0 2.4 1.0 1.0 1.0 2.4 1.0 1.0 1.0 2.4 1.0 1.0 1.0 2.4 0.1 0.1
0.1 0.1 0.1 0.1 0.1 0.1];

% extended low loading transient
%LFD = [0.1 0.1 0.1 0.1 0.1 0.1 0.1 0.1 2.4 2.4 0.1 1.0 2.4 2.4 0.1 1.0 2.4 2.4 0.1 1.0 0.1 0.1
0.1 0.1 0.1 0.1 0.1 0.1];

%extended steady state loading
LFD = [1.0 1.0 1.0 1.0 1.0 1.0 1.0 1.0 1.0 1.0 1.0 1.0 1.0 1.0 1.0 1.0 1.0 1.0 1.0 1.0 1.0 1.0 1.0
1.0 1.0 1.0 1.0 1.0];

%steady state loading
LFSS = [1.0 1.0 1.0 1.0 1.0 1.0 1.0 1.0 1.0 1.0 1.0 1.0 1.0 1.0 1.0 1.0 1.0 1.0 1.0 1.0 1.0 1.0 1.0
1.0 1.0 1.0 1.0 1.0];

%Initial Conditions Matrix
xo = [IntHAc.*LFSS(1,1) IntP intbiomass inttotpha intpp inttotgly intpam intgam intpampha
intgampha intpamgly intgamgly];

%Preallocation of Z Matrix
Z = zeros(17,cycles);

%Logic
for i = 1:cycles

if xo(1,1) < 50
ze1 = 0.54;
ze2 = 0.94;
end
```

```
if xo(1,1) > 50
if xo(1,1) < 130
ze1 = 0.56;
ze2 = 0.80;
end
end

if xo(1,1) > 130
ze1 = 0.58;
ze2 = 0.52;
end


%Solve the differential equations in the BIOP file
[t,x] = ode23s('BIOP30',TSPAN,xo,[],anHRT);

%Reassign the values of the initial conditions matrix
%and assign the final values to table Z
if rem(i,4) == 0
   if rem(i,28) == 0
    hh = hh + 1;
    men = i - 28*hh+1;
 else
    men = i - 28*hh+1;
 end

   if i < brk
   xo = [IntHAc.*LFSS(1,men) IntP j.*x(length(x),3) j.*x(length(x),4) j.*x(length(x),5)
j.*x(length(x),6) j.*x(length(x),7) j.*x(length(x),8) j.*x(length(x),9) j.*x(length(x),10)
j.*x(length(x),11) j.*x(length(x),12)];
else
   xo = [IntHAc.*LFD(1,men) IntP j.*x(length(x),3) j.*x(length(x),4) j.*x(length(x),5)
j.*x(length(x),6) j.*x(length(x),7) j.*x(length(x),8) j.*x(length(x),9) j.*x(length(x),10)
j.*x(length(x),11) j.*x(length(x),12)];
end

 for h=1:17
    if h == 13
       if i < brk
          if rem(i,28) == 0
             Z(h,i) = IntHAc.*LFSS(1,men+27);
          else
             Z(h,i) = IntHAc.*LFSS(1,men-1);
          end

       else
```

```
        if rem(i,28) == 0
          Z(h,i) = IntHAc.*LFD(1,men+27);
        else
          Z(h,i) = IntHAc.*LFD(1,men-1);
        end
      end

    end

   if h<3
     Z(h,i) = x(length(x),h);
   end
   if h>2
   if h<13
     Z(h,i) = j.*x(length(x),h);
   end
   end

   if h == 14
     Z(h,i) = x(21,2)-x(1,2);
   end
   if h == 15
     Z(h,i) = (x(1,1)-x(21,1));
   end
   if h == 16
     Z(h,i) = ((x(21,9)-x(1,9))/1.33);
   end
        if h == 17
     Z(h,i) = x(length(x),5)-x(21,5);
   end
end

else

   men = i - 28*hh+1;

   if i < brk
xo = [IntHAc.*LFSS(1,men) IntP x(length(x),3) x(length(x),4) x(length(x),5) x(length(x),6)
x(length(x),7) x(length(x),8) x(length(x),9) x(length(x),10) x(length(x),11) x(length(x),12)];
else
xo = [IntHAc.*LFD(1,men) IntP x(length(x),3) x(length(x),4) x(length(x),5) x(length(x),6)
x(length(x),7) x(length(x),8) x(length(x),9) x(length(x),10) x(length(x),11) x(length(x),12)];
end

for h=1:17
   if h == 13
```

```
      if i < brk
        Z(h,i) = IntHAc.*LFSS(1,men-1);
      else
        Z(h,i) = IntHAc.*LFD(1,men-1);
      end
    end
    if h<13
      Z(h,i) = x(length(x),h);
    end
          if h == 14
      Z(h,i) = x(21,2)-x(1,2);
    end
    if h == 15
      Z(h,i) = (x(1,1)-x(21,1));
    end
    if h == 16
      Z(h,i) = ((x(21,9)-x(1,9))/1.33);
    end
    if h == 17
      Z(h,i) = x(length(x),5)-x(21,5);
    end
end

end


 %stability check parameters
if i == cycles - 16
  checktot1 = Z(3,i);
  checkpam1 = Z(7,i);
  checkgam1 = Z(8,i);
  checkpampha1 = Z(9,i);
  checkgampha1 = Z(10,i);
  checkpamgly1 = Z(11,i);
  checkgamgly1 = Z(12,i);
  checkpha1 = Z(4,i);
end

if i == cycles
  checktot = Z(3,i);
  checkpam = Z(7,i);
  checkgam = Z(8,i);
  checkpampha = Z(9,i);
  checkgampha = Z(10,i);
  checkpamgly = Z(11,i);
  checkgamgly = Z(12,i);
```

```
    checkpha = Z(4,i);
  end
end

%Preallocation of additional matrices for plotting purposes
S = zeros(2,cycles);
%S is the effluent P, HAc matrix;

W = zeros(3,cycles);
%W is the EOA VSS matrix;
%end

F = zeros(2,cycles);
%F is the EOA PHA and CH matrix;

H = zeros(1,cycles);
%H is the P release matrix;

WILL = zeros(1,cycles);
%WILL is the acetate uptake matrix;

FR = zeros(1,cycles);
%FR is the acetate taken up by PAM;

ZOO = zeros(1,cycles);
%ZOO is the PP synthesized during aerobic phase

PMT = zeros(2,cycles);
%PMT is PAM/GAM PHA

CPT = zeros(2,cycles);
%CPT is PAM/GAM CH

%Assigning values to supplemental matrices
for i = 1:cycles

%Assigning values to supplemental matrix S
S(1,i) = Z(1,i)*30;
S(2,i) = Z(2,i)*31;
%S(1,i) = Z(5,i);

%Assigning values to supplemental matrix W

W(1,i) = (Z(3,i))*128/5 + (Z(4,i))*102/4 + (Z(6,i))*30;
W(2,i) = (Z(7,i))*128/5 + (Z(9,i))*102/4 + (Z(11,i))*30;
W(3,i) = (Z(8,i))*128/5 + (Z(10,i))*102/4 + (Z(12,i))*30;
```

```
%Assigning values to supplemental matrix F

F(1,i) = ((Z(4,i))*102/4)./(W(1,i))*100;
F(2,i) = ((Z(6,i))*30)./(W(1,i))*100;
%F(3,i) = Z(8,i);

%Assigning values to supplemental matrix H

H(1,i) = Z(14,i)*31;
%H(2,i) = Z(11,i);
%H(3,i) = Z(12,i);

%Assigning values to supplemental matrix WILL
WILL(1,i) = Z(15,i)*30;

%Assigning values to supplemental matrix FR
FR(1,i) = Z(16,i)*30;

%Assigning values to supplemental matrix ZOO
ZOO(1,i) = Z(17,i)*31;

%Assigning values to supplemental matrix PMT
PMT(1,i) = Z(9,i)*102/4;
PMT(2,i) = Z(10,i)*102/4;

%Assigning values to supplemental matrix CPT
CPT(1,i) = Z(11,i)*30;
CPT(2,i)= Z(12,i)*30;

end

%Plotting trend figures
I = [1:cycles];

figure(7012)
subplot(2,2,1)
plot(I,S)
title('Effluent P and Acetate')
xlabel('No. Cycles')
ylabel('mg/L')
%axis([56 128 0 100])

subplot(2,2,2)
plot(I,W)
title('EOA VSS')
```

```
xlabel('No. Cycles')
ylabel('mgVSS/L')
%axis([56 128 0 600])

subplot(2,2,3)
plot(I,F)
title('EOA PHA and CH')
xlabel('No. Cycles')
ylabel('% by weight of VSS')
%axis([56 128 0 20])

subplot(2,2,4)
plot(I,H)
title('P Release')
xlabel('No. Cycles')
ylabel('mgP/L')
%axis([56 128 1.0 12.0])

figure(9224)
subplot(2,1,1)
plot(I,WILL)
title('Anaerobic Acetate Uptake')
xlabel('No. Cycles')
ylabel('mgHAc/L')
%axis([56 128 0 150])

subplot(2,1,2)
plot(I,FR)
title('Acetate taken by PAM')
xlabel('No Cycles')
ylabel('mgHAc/L')
%axis([56 128 0 30])

figure(7098)
plot(I,ZOO)
title('PP synthesized')
xlabel('No Cycles')
ylabel('mgPP-P/L')
%axis([56 128 0 20])

figure(1234)
subplot(2,1,1)
plot(I,PMT)
title('Biomass PHA')
xlabel('No. Cycles')
ylabel('mg PHA/L')
```

```
%axis([56 128 0 80])

subplot(2,1,2)
plot(I,CPT)
title('Biomass CH')
xlabel('No. Cycles')
ylabel('mg CH/L')
%axis([56 128 0 120])
```

B.2.3 Reaction Rate Expressions

```
function xdot = BIOP30(t,x,flag,anHRT)
%xdot is the matrix of differential expressions (ie dS/dt, dP/dt,etc...)
%x is the matrix of parameters (ie S, P, etc...)

%The parameters in the x matrix are respectively:
%C-Hac (1) P-PO4 (2) Biomass (3) C-pha (4) PP (5) C-Gly (6)
%C-PAMbiomass (7) C-GAMbiomass (8) C-PAMpha (9) C-GAMpha (10) C-PAMgly (11) C-
GAMgly (12)

%These parameters are in terms of mM C, or mM P (per hour where applicable)

%global declarations
global ico ICPLF pqsmax gqsmax default initconditions count fc pfraction pkx gkx intpp gks
pks pamy gamy IntP10 pfglmax gfglmax pkgl gkgl

%PAM constants
pman = 0.001;
pkpp = 0.535;
pkp = 0.01;
pfppmax = 0.3;
pmaer = 0.004;
cpp = 0.15;
kpgrow = 0.015;
pphaxmax = 0.6;
pphaxmin = 0.00001;
pglyxmin = 0.00001;
ppmin = 0.00000000001;
pneed = 0.01*128/31/5;

%GAM constants
gmaer = 0.001;
gphaxmax = 0.85;
gphaxmin = 0.00001;
gglyxmin = 0.00001;

%aerobic uptake of substrate (HAc)
```

```
paeroqsmax = 0.108;
%paeroqsmax = 0.108;
gaeroqsmax = 0.08;
%aeroqsmax units (mM c-HAc/mM c-biomass-hr)(re:Schuler 98)
paeroks = 0.033;
gaeroks = 0.033;
%aeroks units (mM C-HAc)(re:Schuler 98)
pamy = 0.28;
gamy = 0.28;

%logic
if t<anHRT
  %anaerobic metabolism - kinetic expressions
  xdot = [((-1*pqsmax.*x(1)./(x(1)+pks).*x(7)*(1-ppmin/x(5))-
1*gqsmax.*x(1)./(x(1)+gks).*x(8))*(x(1)>0);
     ((pqsmax.*x(1)./(x(1)+pks).*x(7)*ze1* + pman.*x(7))*(1-ppmin/x(5)))*(x(1)>0);
     0.0;
     ((pqsmax.*x(1)./(x(1)+pks).*x(7)*1.2)*(1-
ppmin/x(5))+gqsmax.*x(1)./(x(1)+gks).*x(8)*1.2;
     (-pqsmax.*x(1)./(x(1)+pks).*x(7)*0.55* - pman.*x(7))*(1-ppmin/x(5))*(x(5)>0);
     ((-pqsmax.*x(1)./(x(1)+pks).*x(7)*0.50)+(-gqsmax.*x(1)./(x(1)+gks).*x(8)*ze2-
gman.*x(8)))*(x(1)>0);
     0.0;
     0.0;
     ((pqsmax.*x(1)./(x(1)+pks).*x(7)*1.2)*(x(1)>0));
     (gqsmax.*x(1)./(x(1)+gks).*x(8)*1.2)*(x(1)>0);
     ((-pqsmax.*x(1)./(x(1)+pks).*x(7)*0.50))*(x(1)>0);
     (-gqsmax.*x(1)./(x(1)+gks).*x(8)*ze2-gman.*x(8))*(x(1)>0)];
else
  %aerobic metabolism - kinetic expressions
  xdot = [((-1*paeroqsmax.*x(1)./(x(1)+paeroks)).*x(7)-
(gaeroqsmax.*x(1)./(x(1)+gaeroks)).*x(8))*(x(1)>0);
     -pneed*pkx.*x(9)-pkpp.*x(2)./(x(2)+pkp).*(1-x(5)./(x(7)*pfppmax)).*x(9)-
pneed*gkx.*x(10);
     pkx.*x(9)+ gkx.*x(10);
     (-1.37*pkx.*x(9)-0.59*pkpp.*x(2)./(x(2)+pkp).*(1-x(5)./(x(7)*pfppmax)).*x(9)-
1.285*pkgl.*(pfglmax-x(11)./x(7)).*x(7)-pmaer.*x(7) +
(paeroqsmax.*x(1)./(x(1)+paeroks)).*x(7))+(-1.37*gkx.*x(10)-gkgl.*(gfglmax-
x(12)./x(8)).*x(8)-gmaer.*x(8) + (gaeroqsmax.*x(1)./(x(1)+gaeroks)).*x(8));
     pkpp.*x(2)./(x(2)+pkp).*(1-x(5)./(x(7)*pfppmax)).*x(9);
     pkgl.*(pfglmax-x(11)./x(7)).*x(7)+ gkgl.*(gfglmax-x(12)./x(8)).*x(8);
     pkx.*x(9);
     gkx.*x(10);
     (-1.37*pkx.*x(9)-0.59*pkpp.*x(2)./(x(2)+pkp).*(1-x(5)./(x(7)*pfppmax)).*x(9)-
1.278*pkgl.*(pfglmax-x(11)./x(7)).*x(7)-pmaer.*x(7) +
(paeroqsmax.*x(1)./(x(1)+paeroks)).*x(7));
```

```
    (-1.37*gkx.*x(10)-gkgl.*(gfglmax-x(12)./x(8)).*x(8)-gmaer.*x(8) +
(gaeroqsmax.*x(1)./(x(1)+gaeroks)).*x(8));
    pkgl.*(pfglmax-x(11)./x(7)).*x(7);
    gkgl.*(gfglmax-x(12)./x(8)).*x(8)];
  end
```

B.3 CA Model Development

B.3.1 Development of Mathematical Expressions and Calibration

The rate expressions in the CA model were similar to those of the GAM, describing the removal of acetate and P, cell synthesis, and the associated transformation of PHA and CH. The rate expressions are shown in Table B-5. Model calibration was conducted by trial and error until model simulations agreed with experimental data. Parameter values and stoichiometry are shown in Tables B-6 and B-7.

Table B-5. CA Model Rate Expressions.

Process	Specific Rate Expression (mole/mole/h) and variables (mM)	Parameters	Note
Substrate Uptake	$q_{smax}*[C_s/(C_s+K_s)]$ C_s = mM Acetate-C	q_{smax} K_s	1
Biomass Synthesis	$k_x\ (C_{pha}/C_{biomass})$ C_{pha} = mM PHA-C $C_{biomass}$ = mM Biomass-C	[3]k_x	2
Glycogen Formation	$k_{gly}*[(C_{glymax}-C_{gly})/C_{biomass}]$ C_{gly} = mM GLY-C $C_{biomass}$ = mM Biomass-C	k_{gly} $C_{glymax} = 0.27*C_{biomass}$	2
Maintenance	constant	m	2

[1] Stoichiometrically linked to PHA synthesis, and CH degradation rate expressions
[2] Stoichiometrically linked to PHA degradation rate expression
[3] Target Parameter

Table B-6. CA Model Parameter Values1.

Parameter	Value	Unit
k_x	0.1	mol C/mol C/h
q_{smax}	0.15	mol C/mol C/h
K_s	1.6	mol C/L
k_{gl}	0.15	mol C/mol C/h
p_m	0.004	mol C/mol C/h
[1]Current Research		

Table B-7. CA Model Stoichiometry.

Ratio	Coefficient	Reference
PHA-C Synthesized/Acetate-C Removed	variable[1]	2
GLY-C Degraded/Acetate-C Removed	variable[1]	2
PHA-C Degraded/ GLY-C Synthesized	0.25	2
[1]A function of Influent COD Peak Load Factor (See section 3.4.3.) [2]Current Research		

B.3.2 Specification of Parameters

```
%This program calls the goa program
tstart = clock;

%global declarations
global pqsmax gqsmax default initconditions count fc pfraction pkx gkx ze3 ze4 gks pks ICPLF ico
global checktot1 checkpam1 checkgam1 checktot checkpam checkgam
global aaa bbb ccc ddd eee fff ggg hhh iii jjj kkk lll pamy gamy IntP10 pfglmax gfglmax pkgl gkgl

%value assignments
tolerance = 0.5;
default = 1;
initconditions = 2;
countt = 1;
kx = 0.10;
qsmax = 0.15;
ks = 1.6;
```

```
gamy = 0.28;
IntP10 = 1.1;
fglmax = 0.14;
kgl = 0.15;
counter = 9;
stable = 0;

goa

tstop = clock;
runtime = etime(tstop,tstart);
runtime = runtime/60;
disp(runtime)
disp('minutes')
%end
```

B.3.3 Reactor Operation and Graphing

```
%This file calculates the CA profile sequentially

%The parameters in the matrix are respectively:
%C-HAc (1) P-PO4 (2) Biomass (3) C-PHA (4) PP (5) C-Gly

%process parameters
hh = 0;
%I've added the hh parameter to count the number of times that wasting occurs in order to help
set the
%loading index called "men" men uses the cycles index i and hh to set the value of the influent
COD loading

poneHRT = 0.85;
aerHRT = 3.95;
MCRT = 4;
jj = 1/MCRT;
j = 1 - jj;
IntHAc10 = 94;
%IntHAc10 is initial HAc as mg/l HAc
IntHAc = IntHAc10/30;
%IntP10 = 1;
%IntP10 is initial P as mg/l P
IntP = IntP10/31;
Trans = 1;
cycles = 64;
brk = 84;
```

```
%Time interval
to = 0;
tf = poneHRT+aerHRT;
t1 = poneHRT;
t2 = to + .4;
t3 = t2 + .4;
t4 = t3 + .4;
t5 = t4 + .4;
t6 = t5 + .4;
t7 = t6 + .4;
t8 = t7 + .4;
t9 = t8 + .4;
t10 = t9 + .4;
TSPAN = [to tf];

%Initial Conditions
intbiomass = 20;
%Initial intbiomass = 20;
inttotpha = 0.2;
inttotgly = 0.5;

%high/low combination transient
%LFD = [0.2 1.0 1.0 1.8 0.2 1.0 1.0 1.8 0.2 1.0 1.0 1.8 0.2 1.0 1.0 1.8 0.2 1.0 1.0 1.8 0.2 1.0
1.0 1.8 0.2 1.0 1.0 1.8];

%weekend low loading transient
%LFD = [1.0 1.0 1.0 2.4 1.0 1.0 1.0 2.4 1.0 1.0 1.0 2.4 1.0 1.0 1.0 2.4 1.0 1.0 1.0 2.4 0.1 0.1
0.1 0.1 0.1 0.1 0.1 0.1];

% extended low loading transient
%LFD = [0.1 0.1 0.1 0.1 0.1 0.1 0.1 0.1 2.4 2.4 0.1 1.0 2.4 2.4 0.1 1.0 2.4 2.4 0.1 1.0 0.1 0.1
0.1 0.1 0.1 0.1 0.1 0.1];

%extended steady state loading
 LFD = [1.0 1.0 1.0 1.0 1.0 1.0 1.0 1.0 1.0 1.0 1.0 1.0 1.0 1.0 1.0 1.0 1.0 1.0 1.0 1.0 1.0 1.0 1.0
1.0 1.0 1.0 1.0 1.0];

%steady state loading
LFSS = [1.0 1.0 1.0 1.0 1.0 1.0 1.0 1.0 1.0 1.0 1.0 1.0 1.0 1.0 1.0 1.0 1.0 1.0 1.0 1.0 1.0 1.0 1.0
1.0 1.0 1.0 1.0 1.0];

%Initial Conditions Matrix
xo = [IntHAc.*LFSS(1,1) IntP intbiomass inttotpha inttotgly];

%Preallocation of Z Matrix
Z = zeros(17,cycles);
```

```
%Logic
for i = 1:cycles

if xo(1,1) < 50
ze3 = 0.33;
ze4 = 0.28;
end

if xo(1,1) > 50
if xo(1,1) < 130
ze3 = 0.35;
ze4 = 0.30;
end
end

if xo(1,1) > 130
ze3 = 0.61;
ze4 = 0.43;
end

%Solve the differential equations in the BIOP file
[t,x] = ode23s('ca',TSPAN,xo,[],poneHRT);

%Reassign the values of the initial conditions matrix
%and assign the final values to table Z
if rem(i,4) == 0
   if rem(i,28) == 0
    hh = hh + 1;
    men = i - 28*hh+1;
 else
    men = i - 28*hh+1;
 end

   if i < brk
   xo = [IntHAc.*LFSS(1,men) IntP j.*x(length(x),3) j.*x(length(x),4) j.*x(length(x),5)];
else
   xo = [IntHAc.*LFD(1,men) IntP j.*x(length(x),3) j.*x(length(x),4) j.*x(length(x),5)];
end

 for h=1:5
      if i < brk
         if rem(i,28) == 0
            Z(h,i) = IntHAc.*LFSS(1,men+27);
         else
            Z(h,i) = IntHAc.*LFSS(1,men-1);
```

```
        end

      else
        if rem(i,28) == 0
          Z(h,i) = IntHAc.*LFD(1,men+27);
        else
          Z(h,i) = IntHAc.*LFD(1,men-1);
        end
     end

  if h<3
     Z(h,i) = x(length(x),h);
  else
     Z(h,i) = j.*x(length(x),h);
  end

end

else

  men = i - 28*hh+1;

  if i < brk
xo = [IntHAc.*LFSS(1,men) IntP x(length(x),3) x(length(x),4) x(length(x),5)];
else
xo = [IntHAc.*LFD(1,men) IntP x(length(x),3) x(length(x),4) x(length(x),5)];
end

for h=1:5
  if i < brk
       Z(h,i) = IntHAc.*LFSS(1,men-1);
  else
       Z(h,i) = IntHAc.*LFD(1,men-1);
  end
     Z(h,i) = x(length(x),h);

end

end
end

%Preallocation of additional matrices for plotting purposes
S = zeros(2,cycles);
%S is the effluent P, HAc matrix;

W = zeros(2,cycles);
```

```
%W is the EOA VSS matrix;
%end

F = zeros(1,cycles);
%F is the EOA PHA and CH matrix;

H = zeros(1,cycles);
%WILL is the acetate uptake matrix;

%Assigning values to supplemental matrices
for i = 1:cycles

%Assigning values to supplemental matrix S
S(1,i) = Z(1,i)*30;
S(2,i) = Z(2,i)*31;
%S(1,i) = Z(5,i);

%Assigning values to supplemental matrix W

W(1,i) = (Z(3,i))*128/5 + (Z(4,i))*102/4 + (Z(5,i))*30;
W(2,i) = (Z(3,i))*128/5;
%W(3,i) = (Z(8,i))*128/5 + (Z(10,i))*102/4 + (Z(12,i))*30;

%Assigning values to supplemental matrix F

F(1,i) = ((Z(4,i))*102/4)./(W(1,i))*100;
%F(2,i) = ((Z(6,i))*30)./(W(1,i))*100;
%F(3,i) = Z(8,i);

%Assigning values to supplemental matrix H

H(1,i) = Z(5,i)*30./(W(1,i))*100;
%H(2,i) = Z(11,i);
%H(3,i) = Z(12,i);

end

%Plotting trend figures
I = [1:cycles];

figure(7012)
subplot(2,2,1)
plot(I,S)
title('Effluent P and Acetate')
xlabel('No. Cycles')
ylabel('mg/L')
```

```
subplot(2,2,2)
plot(I,W)
title('EOA VSS and Biomass VSS')
xlabel('No. Cycles')
ylabel('mg VSS/L')

subplot(2,2,3)
plot(I,F)
title('EOA PHA')
xlabel('No. Cycles')
ylabel('% by weight of VSS')

subplot(2,2,4)
plot(I,H)
title('EOA CH')
xlabel('No. Cycles')
ylabel('% by weight of VSS')

figure(500)
subplot(2,1,1)
plot(t,x(:,[1 2]))
title('Acetate and P Time Series')
xlabel('time')
ylabel('mM C')

subplot(2,1,2)
plot(t,x(:,[4 5]))
title('Biomass Time Series')
xlabel('time')
ylabel('mM C')
```

B.3.4 Reaction Rate Expressions

```
function xdot = CA(t,x,flag,poneHRT)
%xdot is the matrix of differential expressions (ie dS/dt, dP/dt,etc...)
%x is the matrix of parameters (ie S, P, etc...)

%The parameters in the x matrix are respectively:
%C-Hac (1) P-PO4 (2) Biomass (3) C-pha (4) PP (5) C-Gly

%These parameters are in terms of mM C, or mM P (per hour where applicable)

%global declarations
global ico ICPLF pqsmax gqsmax default initconditions count fc pfraction kx intpp ks gamy
IntP10 fglmax kgl
```

```
%constants
%ks = 1.0;
%man = 0.001;
%kgl = 0.17;
%kgl originally 0.6, 0.2 gives me negative glycogen values
%fglmax = 0.18;
maer = 0.001;
phaxmax = 0.85;
phaxmin = 0.00001;
glyxmin = 0.00001;



%aerobic uptake of substrate (HAc)
aeroqsmax = 0.10;

%aeroks units (mM C-HAc)(re:Schuler 98)
aeroks = 0.033;

gamy = 0.28;

%logic
if t<poneHRT
  %phase one metabolism - kinetic expressions
  xdot = [(-1*qsmax.*x(1)./(x(1)+ks).*x(3))*(x(1)>0);
    -pneed*kx.*x(4);
    gkx.*x(4);
    -0.84*gkx.*x(4) + qsmax.*x(1)./(x(1)+gks).*x(3)*ze3;
    -qsmax.*x(1)./(x(1)+gks).*x(3)*ze4-man.*x(3)];
else
  %phase two metabolism - kinetic expressions
  xdot = [(-1*qsmax.*x(1)./(x(1)+ks).*x(3))*(x(1)>0);
    -pneed*kx.*x(4);
    kx.*x(4);
    (-0.84*kx.*x(4)-kgl.*(fglmax-x(5)./x(3)).*x(3)-maer.*x(3) +
0.40*(aeroqsmax.*x(1)./(x(1)+aeroks)).*x(3));
    kgl.*(fglmax-x(5)./x(3)).*x(3) - 0.25*0.40*(aeroqsmax.*x(1)./(x(1)+aeroks)).*x(3)];
end
```

REFERENCES

Arun, V., Mino, T., and Matsuo, T. 1988. Biological mechanism of acetate uptake mediated by carbohydrate consumption in excess phosphorus removal systems. *Water Res.* 22(5): 565.

Burkhead, C., and McKinney, R. 1969. Energy concepts of aerobic microbial metabolism. *J. Sanit. Eng*. 95: 253

Cech, J., and Hartman, P. 1993. Competition between polyphosphate and polysaaccharide-accumulating bacteria in enhanced biological phosphorus removal. *Water Res.* 27(7): 1219.

Filipe, C., and Daigger, G. 1998. Development of a revised metabolic model for the growth of phosphorus-accumulating organisms. *Water Environ. Res*. 70(1): 67.

Helmers, E., Frame, J., Greenberg, A., and Sawyer, C. 1951. Nutritional requirements in the biological stabilization of industrial wastes: Part II. Treatment with domestic sewage. *Sewage and Indust. Wastes* 23(7): 884.

Jenkins, D., Richard, M., and Daigger, G. 1993. *Manual on the Causes and Control of Activated Sludge Bulking and Foaming.* Boca Raton, FL: Lewis Publishers.

Jones, P. 1965. The Effect of Nitrogen and Phosphorus Compounds on One of the Microrganisms Responsible for Sludge Bulking. *Proc. 7th Industrial Waste Conference,* Purdue University.

Kustu, S., McFarland, N., Hui, S., Esmon, B., and Ames, G. 1979. Nitrogen control in Salmonella typhimurium: co-regulation of synthesis of glutamine synthetase and amino-acid transport systems. *J. Bacteriol.* 138(1): 218.

Lehninger, A., Nelson, D., and Cox, M. (1993). *Principles of Biochemistry,* Second Edition, New York, NY: Worth Publishers.

Liu, W., Nakamura, K., Matsuo, T., and Mino, T. 1997. Internal energy-based competition between polyphosphate- and glycogen-accumulating bacteria in biological phosphorus removal reactors - effect of P/C feeding ratio. *Water Res.* 31(6): 1430.

Metcalf and Eddy, Inc. 1991. *Wastewater Engineering: Treatment, Disposal, and Reuse,* Third Edition. New, York, NY: McGraw-Hill Publishers.

Mino, T., van Loosdrecht, M., and Hiejnen, J. 1998. Microbiology and biochemistry of the enhanced biological phosphate removal process. *Water Res.* 32(11): 3193.

Satoh, H., Mino, T., and Matsuo, T. 1992. Uptake of organic substrates and accumulation of polyhydroxyalkanoates linked with glycolysis of intracellular carbohydrates under anaerobic conditions in the biological excess phosphate removal processes. *Water Sci. and Technol.* 26: 933.

Sawyer, C. 1941. B.O.D. removal from waste sulfite liquor-sewage mixtures by activated sludge—factors governing the rate of removal. *Indust. Eng. Chem.* 33: 411.

Schuler, A. 1998. *The Effects of Varying Influent Phosphate and Acetate Concentrations on Enhanced Biological Removal of Phosphate from Wastewater*. Ph.D. Dissertation, Civil and Environ. Eng. University of California, Berkeley.

Smolders, G., van der Meij, J., van Loosdrecht, M., and Hiejnen, J. 1995. A structured metabolic model of the anaerobic and aerobic stoichiometry and kinetics of the biological phosphorus removal process. *Biotechnol. and Bioeng.* 47: 277.

WASTEWATER UTILITY

Alabama
Montgomery Water Works & Sanitary Sewer Board

Alaska
Anchorage Water & Wastewater Utility

Arizona
Gila Resources
Glendale, City of, Utilities Department
Mesa, City of
Phoenix Water Services Department
Pima County Wastewater Management

Arkansas
Little Rock Wastewater Utility

California
Calaveras County Water District
Central Contra Costa Sanitary District
Contra Costa Water District
Crestline Sanitation District
Delta Diablo Sanitation District
Dublin San Ramon Services District
East Bay Municipal Utility District
El Dorado Irrigation District
Escondido, City of
Fairfield-Suisun Sewer District
Irvine Ranch Water District
Las Virgenes Municipal Water District
Lodi, City of
Los Angeles, City of
Los Angeles County, Sanitation Districts of
Napa Sanitation District
Orange County Sanitation District
Palo Alto, City of
Riverside, City of
Sacramento Regional County Sanitation District
San Diego Metropolitan Wastewater Department, City of
San Francisco, City & County of
San Jose, City of
Santa Barbara, City of
Santa Rosa, City of
South Bayside System Authority
South Orange County Wastewater Authority
Union Sanitary District
West Valley Sanitation District

Colorado
Boulder, City of
Colorado Springs, City of
Littleton/Englewood Water Pollution Control Plant
Metro Wastewater Reclamation District, Denver

Connecticut
New Haven, City of, WPCA

District of Columbia
District of Columbia Water & Sewer Authority

Florida
Broward, County of
Fort Lauderdale, City of
Gainesville Regional Utilities
JEA
Kissimmee, City of, Department of Water Resources
Miami-Dade Water & Sewer Authority
Orange County Utilities Department
Orlando, City of
Reedy Creek Improvement District
Seminole County Environmental Services
St. Petersburg, City of
Stuart Public Utilities
Tallahassee, City of
Tampa, City of
West Palm Beach, City of

Georgia
Atlanta Department of Watershed Management
Augusta, City of
Clayton County Water Authority
Cobb County Water System
Columbus Water Works
Fulton County
Gwinnett County Department of Public Utilities
Macon Water Authority
Savannah, City of

Hawaii
Honolulu, City and County of

Idaho
Pocatello, City of

Illinois
American Bottoms Wastewater Treatment Plant
Dupage, County of
Greater Peoria Sanitary District
Kankakee River Metropolitan Agency
Metropolitan Water Reclamation District of Greater Chicago
North Shore Sanitary District
Wheaton Sanitary District

Indiana
Fort Wayne, City of

Iowa
Ames, City of
Des Moines Metro Wastewater Reclamation Authority

Kansas
Johnson County Unified Wastewater Districts
Unified Government of Wyandotte County/Kansas City, City of

Kentucky
Louisville & Jefferson County Metropolitan Sewer District

Louisiana
Sewerage & Water Board of New Orleans

Maine
Bangor, City of
Portland Water District

Maryland
Anne Arundel County Bureau of Utility Operations
Howard County Department of Public Works
Washington Suburban Sanitary Commission

Massachusetts
Boston Water & Sewer Commission
Upper Blackstone Water Pollution Abatement District

Michigan
Ann Arbor, City of
Detroit, City of
Grand Rapids, City of
Holland Board of Public Works
Lansing, City of
Owosso Mid-Shiawassee County WWTP
Saginaw, City of
Wayne County Department of Environment
Wyoming, City of

Minnesota
Rochester, City of

Missouri
Independence, City of
Kansas City Missouri Water Services Department
Little Blue Valley Sewer District
Metropolitan St. Louis Sewer District

Nebraska
Lincoln Wastewater System

Nevada
Henderson, City of

New Jersey
Passaic Valley Sewerage Commissioners

New York
New York City Department of Environmental Protection

North Carolina
Cary, Town of
Charlotte/Mecklenburg Utilities
Durham, City of
Metropolitan Sewerage District of Buncombe County
Orange Water & Sewer Authority

Ohio
Akron, City of
Butler County Department of Environmental Services
Columbus, City of
Metropolitan Sewer District of Greater Cincinnati
Northeast Ohio Regional Sewer District
Summit, County of

Oklahoma
Tulsa, City of

Oregon
Clean Water Services
Eugene/Springfield Water Pollution Control
Water Environment Services

Pennsylvania
Philadelphia, City of
University Area Joint Authority, State College

South Carolina
Charleston Commissioners of Public Works
Mount Pleasant Waterworks & Sewer Commission
Spartanburg Sanitary Sewer District

WERF SUBSCRIBERS

Tennessee
Cleveland, City of
Knoxville Utilities Board
Murfreesboro Water & Sewer Department
Nashville Metro Water Services

Texas
Amarillo, City of
Austin, City of
Dallas Water Utilities
Denton, City of
El Paso Water Utilities
Fort Worth, City of
Gulf Coast Waste Disposal Authority
Houston, City of
San Antonio Water System
Trinity River Authority

Utah
Salt Lake City Corporation

Virginia
Alexandria Sanitation Authority
Arlington, County of
Fairfax County Virginia
Hampton Roads Sanitation District
Henrico, County of
Hopewell Regional Wastewater Treatment Facility
Loudoun County Sanitation Authority
Lynchburg Regional WWTP
Prince William County Service Authority
Richmond, City of

Washington
Edmonds, City of
Everett, City of
King County Department of Natural Resources
Seattle Public Utilities
Yakima, City of

Wisconsin
Green Bay Metro Sewerage District
Madison Metropolitan Sewerage District
Milwaukee Metropolitan Sewerage District
Racine, City of
Sheboygan Regional Wastewater Treatment
Wausau Water Works

Australia
South Australian Water Corp.
Sydney Water Corp.
Water Corp. of Western Australia

Canada
Greater Vancouver Regional District
Toronto, City of, Ontario

Mexico
Servicios de Agua y Drenaje de Monterrey, I.P.D.

United Kingdom
Yorkshire Water Services Limited

STORMWATER UTLITY

California
Los Angeles, City of, Department of Public Works
Monterey, City of
Santa Rosa, City of

Colorado
Boulder, City of

Georgia
Griffin, City of

Kansas
Overland Park, City of

Kentucky
Louisville & Jefferson County Metropolitan Sewer District

Maine
Portland Water District

North Carolina
Cary, Town of
Charlotte, City of, Stormwater Services

Pennsylvania
Philadelphia, City of

STATE

Fresno Metropolitan Flood Control District, Calif.
Urban Drainage & Flood Control District, Colo.

CORPORATE

ADS Environmental Services
The ADVENT Group Inc.
Alan Plummer & Associates
Alden Research Laboratory
Alpine Technology Inc.
Aquateam–Norwegian Water Technology Centre A/S
BaySaver Inc.
BioVir Inc.
Black & Veatch
Boyle Engineering Corporation
BPR CSO
Brown & Caldwell
Burns & McDonnell
CABE Associates Inc.
The Cadmus Group
Camp Dresser & McKee Inc.
Carollo Engineers Inc.
Carpenter Environmental Associates Inc.
CDS Technologies Inc.
Chemtrac Systems Inc.
CH2M HILL
Clancy Environmental Consultants Inc.
Damon S. Williams Associates, LLC
David L. Sheridan, P.C.
Earth Tech Inc.
Ecolab Water Care Services
EMA Inc.
The Eshelman Company Inc.
Finkbeiner, Pettis, & Strout Inc.
Frontier Geosciences Inc.
ftn Associates Inc.
Gannett Fleming Inc.
GE Betz
Golder Associates Inc.
Greeley and Hansen LLC
The HACH Company
Hazen & Sawyer, P.C.
HDR Engineering Inc.
HNTB Corporation
HydroQual Inc.
Insituform Technologies Inc.
Institute for Environmental Technology & Industry, Korea
Jacobson Helgoth Consultants Inc.
Jason Consultants Inc.
Jordan, Jones, & Goulding Inc.
KCI Technologies Inc.
Kelly & Weaver, P.C.
Kennedy/Jenks Consultants
Komline Sanderson Engineering Corporation
Lawler, Matusky & Skelly Engineers, LLP
Limno-Tech Inc.
Lombardo Associates Inc.
Malcolm Pirnie Inc.
McKim & Creed
MEC Analytical Systems Inc.
MWH
New England Organics
Odor & Corrosion Technology Consultants Inc. (OCTC)
ONDEO Degremont Inc.
Oswald Green, LLC
PA Government Services Inc.
Parametrix Inc.
Parsons
Post, Buckley, Schuh & Jernigan
R&D Engineering/Conestoga Rover & Associates
The RETEC Group
R.M. Towill Corporation
Ross & Associates Ltd.
Royce Technologies
SAIC Maritime Technical Group
Sear-Brown
Sierra Environmental Services
SYNAGRO
Tetra Tech Inc.
Trojan Technologies Inc.
URS Corporation
USFilter NATC
Vortechnics Inc.
Wade-Trim Inc.
Weston Solutions Inc.
Woodard & Curran
WRc/D&B, LLC
WWETCO, LLC

INDUSTRY

American Electric Power
ChevronTexaco Energy Research & Technology Company
The Coca-Cola Company
Dow Chemical Company
DuPont Company
Eastman Kodak Company
Merck & Company Inc.
ONDEO Services
Procter & Gamble Company
PSEG Services Corp.
Severn Trent Services Inc.
Shell Global Solutions
Thames Water Plc
United Water Services LLC

Note: List as of 5/12/03

WERF Product Order Form

As a benefit of joining the Water Environment Research Foundation, subscribers are entitled to receive one complimentary copy of all reports and other products.Additional copies are available at cost (usually $10).To order your complimentary copy of a report, please w "free" in the unit price column.WERF keeps track of all orders.If the charge differs from what is shown here, we will call to confirm the before processing.

Name Title

Organization

Address

City State Zip Code Country

Phone Fax Email

Stock #	Product	Quantity	Unit Price	Total

Method of Payment:

(All orders must be prepaid.)

❑ Check or Money Order Enclosed

❑ Visa ❑ Mastercard ❑ American Express

Account No. Exp. Date

Signature

Postage & Handling

VA Residents Add 4.5% Sales Tax

Canadian Residents Add 7% GST

TOTAL

Shipping & Handling:

Amount of Order	United States	Canada & Mexico	All Others
Up to but not more than:	Add:	Add:	Add:
$20.00	$5.00*	$8.00	50% of amount
30.00	5.50	8.00	40% of amount
40.00	6.00	8.00	
50.00	6.50	14.00	
60.00	7.00	14.00	
80.00	8.00	14.00	
100.00	10.00	21.00	
150.00	12.50	28.00	
200.00	15.00	35.00	
More than $200.00	Add 20% of order	Add 20% of order	

*minimum amount for all orders

Note: Please make checks payable to the Water Environment Research Foundation.

To Order (Subscribers Only):

Log on to www.werf.org and clic on the "Product Catalog."

Phone: (703) 684-2470
Fax: (703) 299-0742.

WERF
Attn: Subscriber Services
635 Slaters Lane
Alexandria, VA 22314-1177

To Order (Non-Subscribers):

Non-subscribers may be able to orde WERF publications either through WEF (www.wef.org) or IWAP (www.iwapublishing.com).Visit WERI website at www.werf.org for details.

IWA Publishing's authorised EU representative for General Product Safety Regulations is Diane D'Arras, 15 rue Duret, 75116 Paris, France, e-mail: safety@iwap.co.uk.

Printed and bound by CPI Group (UK) Ltd, Croydon, CR0 4YY
12/05/2026
02108881-0014